CAMBRIDGE LIBRARY COLLECTION

Books of enduring scholarly value

Life Sciences

Until the nineteenth century, the various subjects now known as the life sciences were regarded either as arcane studies which had little impact on ordinary daily life, or as a genteel hobby for the leisured classes. The increasing academic rigour and systematisation brought to the study of botany, zoology and other disciplines, and their adoption in university curricula, are reflected in the books reissued in this series.

The Natural History of Birds

Georges-Louis Leclerc, Comte de Buffon (1707–88), was a French mathematician who was considered one of the leading naturalists of the Enlightenment. An acquaintance of Voltaire and other intellectuals, he work as Keeper at the Jardin du Roi from 1739, and this inspired him to research and publish a vast encyclopaedia and survey of natural history, the ground-breaking *Histoire Naturelle*, which he published in forty-four volumes between 1749 and 1804. These volumes, first published between 1770 and 1783 and translated into English in 1793, contain Buffon's survey and descriptions of birds from the *Histoire Naturelle*. Based on recorded observations of birds both in France and in other countries, these volumes provide detailed descriptions of various bird species, their habitats and behaviours and were the first publications to present a comprehensive account of eighteenth-century ornithology. Volume 5 covers larks, wagtails and fig-eaters.

Cambridge University Press has long been a pioneer in the reissuing of out-of-print titles from its own backlist, producing digital reprints of books that are still sought after by scholars and students but could not be reprinted economically using traditional technology. The Cambridge Library Collection extends this activity to a wider range of books which are still of importance to researchers and professionals, either for the source material they contain, or as landmarks in the history of their academic discipline.

Drawing from the world-renowned collections in the Cambridge University Library, and guided by the advice of experts in each subject area, Cambridge University Press is using state-of-the-art scanning machines in its own Printing House to capture the content of each book selected for inclusion. The files are processed to give a consistently clear, crisp image, and the books finished to the high quality standard for which the Press is recognised around the world. The latest print-on-demand technology ensures that the books will remain available indefinitely, and that orders for single or multiple copies can quickly be supplied.

The Cambridge Library Collection will bring back to life books of enduring scholarly value (including out-of-copyright works originally issued by other publishers) across a wide range of disciplines in the humanities and social sciences and in science and technology.

The Natural History of Birds

From the French of the Count de Buffon

Volume 5

Comte de Buffon
William Smellie

CAMBRIDGE UNIVERSITY PRESS

Cambridge, New York, Melbourne, Madrid, Cape Town, Singapore,
São Paolo, Delhi, Dubai, Tokyo, Mexico City

Published in the United States of America by Cambridge University Press, New York

www.cambridge.org
Information on this title: www.cambridge.org/9781108023023

This edition first published 1793
This digitally printed version 2010

ISBN 978-1-108-02302-3 Paperback

THE

NATURAL HISTORY

OF

BIRDS.

FROM THE FRENCH OF THE

COUNT DE BUFFON.

ILLUSTRATED WITH ENGRAVINGS;

AND A

PREFACE, NOTES, AND ADDITIONS,

BY THE TRANSLATOR.

IN NINE VOLUMES.

VOL. V.

LONDON:

PRINTED FOR A. STRAHAN, AND T. CADELL IN THE STRAND;
AND J. MURRAY, Nº 32, FLEET-STREET.

MDCCXCIII.

CONTENTS

OF THE

FIFTH VOLUME.

Page

THE Sky-Lark — — 1

VARIETIES *of the Sky-Lark* — 18

1. The White Sky-Lark — — ib.
2. The Black Sky-Lark — — 20

The Rufous-Backed Lark — — 21

The Wood-Lark — — 23

The Tit-Lark — — 28

VARIETY *of the Tit-Lark* — 32

The White Tit-Lark — — ib.

FOREIGN BIRD *related to the Tit-Lark* — 34

The Louisiana Lark — — ib.

The Grasshopper-Lark — — 36

The Willow-Lark — — 40

The Meadow-Lark — — 41

The Italian Lark — — 45

The Calandre, or Large Lark — 47

FOREIGN BIRDS *analogous to the Calandre-Lark* 51

1. The Cape-Lark — — ib.

CONTENTS.

Page

2. The Shore-Lark — — 53
3. The Brown-cheeked Pennsylvanian Lark 55

The Marsh-Lark — — 57
The Siberian Lark — — 59
FOREIGN BIRDS *which are related to the Lark* 61
1. The Rufous Lark — — ib.
2. The Cinereous Lark — — 62
3. The African Lark — — 63

The Crested Lark — — 65
The Lesser Crested Lark — — 72
The Undated Lark — — 74
FOREIGN BIRD *related to the crested Lark* — 76
The Senegal Lark — — ib.

The Nightingale — — 78
VARIETIES *of the Nightingale* — 105
1. The Great Nightingale — — ib.
2. The White Nightingale — 106

FOREIGN BIRD *related to the Nightingale* 108
The Foudi-Jala — — ib.

SPECIES *of the Fauvette* — — 110
1. The Pettychaps — — 111
2. The Passerinette, or Little Fauvette — 117
3. The Black-Headed Fauvette — 119
4. The Grisette or the Gray Fauvette, called in Provence, Passerine — — 125
5. The Babbler Fauvette — — 128
6. The Russet, or Fauvette of the Woods 131
7. The Reed Fauvette — — 134
8. The Little Rufous Fauvette — 137
9. The Spotted Fauvette — — 140

10. The

CONTENTS.

Page

10. The Winter Fauvette, or Traine-Buiſſon, or Mouchet — — 142
The Alpine Fauvette — — 146
The Pitchou — — 149

FOREIGN BIRDS *related to the Fauvettes* 151
1. The Spotted Fauvette, from the Cape of Good Hope — — — ib.
2. The Small Spotted Fauvette, from the Cape of Good Hope — — 152
3. The Spotted Fauvette, from Louiſiana ib.
4. The Yellow-Breaſted Fauvette, from Louiſiana 154
5. The Rufous-Tailed Fauvette, from Cayenne 155
6. The Fauvette of Cayenne, with a Brown Throat and Yellow Belly — — ib.
7. The Blueiſh Fauvette of Saint Domingo 156

The Yellow-Neck — — — 158
The Redſtart — — — 163
The Red-Tail — — — 171
The Guiana Red-Tail — — 176
The Epicurean Warbler — — 177
The Fiſt of Provence — — 184
The Ortolan Pivote — — ib.
The Red-Breaſt — — — 185
The Blue-Throat — — — 195
FOREIGN BIRD *which is related to the Red-Breaſt and Blue-Throat* — — 200
The Blue Red-Breaſt of North America ib.

The Stone-Chat — — — 203
The Whin-Chat — — — 212
FOREIGN BIRDS *which are related to the Stone-Chat and Whin-Chat* — — — 216
1. The Senegal Stone-Chat or Whin-Chat ib.

2. The

CONTENTS.

Page

2. The Stone-Chat from the Iſland of Luçon — 217
3. Stone-Chat of the Philippines — 218
4. The Great Stone-Chat of the Philippines — 219
5. The Fitert of Madagaſcar Stone-Chat — 220
6. The Great Stone-Chat — — 222
7. The Stone-Chat of the Cape of Good Hope — 223
8. The Spectacle Warbler — 225

The Wheat-Ear — — 228
SPECIES *of the Wheat-Ear* — 234
1. The Wheat-Ear — — ib.
2. The Gray Wheat-Ear — — ib.
3. The Cinereous Wheat-Ear — 235
The Ruſty Wheat-Ear — ib.
The Rufous Wheat-Ear — 236

FOREIGN BIRDS *which are related to the Wheat-Ear* 238
1. The Great Wheat-Ear, or White-Tail of the Cape of Good Hope — — ib.
2. The Greeniſh Brown Wheat-Ear — 239
3. The Senegal Wheat-Ear — 240

The Wagtails — — 241
The White Wagtail — — 242
The Bergeronettes, or Bergerettes — 252
SPECIES *of the Bergeronettes* — ib.
1. The Gray Bergeronette — — ib.
2. The Spring Bergeronette — 256
3. The Yellow Bergeronette — 259

FOREIGN BIRDS *which are related to the Bergeronettes* — — 264
1. The Bergeronette from the Cape of Good Hope ib.
2. The Little Bergeronette from the Cape of Good Hope — — 265
3. The Bergeronette of the Iſland of Timor 266
4. The Bergeronette from Madras — 267

The

CONTENTS.

	Page
The Fig-Eaters — —	269
SPECIES *of the Fig-Eaters* — —	270
1. Green and Yellow Fig-Eater —	ib.
2. The Cheric — —	271
3. The Little Simon — —	273
4. The Blue Fig-Eater —	275
5. The Senegal Fig-Eater — —	276
SPECIES *of the American Fig-Eater*	279
1. The Spotted Fig-Eater —	ib.
2. The Red-Headed Fig-Eater —	280
3. The White-Throated Fig-Eater —	281
4. The Yellow-Throated Fig-Eater —	282
5. The Green and White Fig-Eater —	284
6. The Orange-Throated Fig-Eater —	285
7. The Cinereous-Headed Fig-Eater —	286
8. The Brown Fig-Eater —	287
9. The Black-Cheeked Fig-Eater —	288
10. The Yellow Spotted Fig-Eater —	289
11. The Brown and Yellow Fig-Eater	290
12. The Pine Fig-Eater — —	292
13. The Black-Collared Fig-Eater —	294
14. The Yellow-Headed Fig-Eater —	295
15. The Yellow-Throated Cinereous Fig-Eater	297
16. The Collared Cinereous Fig-Eater —	298
17. The Belted Fig-Eater — —	300
18. The Blue Fig-Eater — —	301
19. The Variegated Fig-Eater —	303
20. The Rufous-Headed Fig-Eater —	304
21. The Red-Breasted Fig-Eater —	306
22. The Cærulean Fig-Eater —	307
23. The Golden-Winged Fig-Eater —	309
24. The Golden-Crowned Fig-Eater —	310
25. The Orange Fig-Eater — —	311
26. The Crested Fig-Eater —	312
27. The Black Fig-Eater — —	313
28. The Olive Fig-Eater — —	314

29. The

CONTENTS.

Page

29. The Prothonotary Fig-Eater — 315
30. The Half-Collared Fig-Eater — 316
31. The Yellow-Throated Fig-Eater 318
32. The Olive-Brown Fig-Eater — 319
33 The Graffet Fig-Eater — — 320
34. The Afh-Throated Cinereous Fig-Eater 321
35. The Great Fig-Eater of Jamaica — 322

The Middle-Bills — — 325
The Worm-Eater — — 327
The Black and Blue Middle-Bill — 329
The Black and Rufous Middle-Bill — 331
The Bimbelé, or Baftard Linnet — 333
The Banana Warbler — — 336
The Middle-Bill, with White Creft and Throat 339
The Simple Warbler — — 341
The Pitpits — — 342
SPECIES *of the Pitpits* — 343
1. The Green Pitpit — ib.
2. The Blue Pitpit — — 344

VARIETIES *of the Blue Pitpit* — ib.
1. The Blue Manakin — — ib.
2. The Blue Pitpit of Cayenne — 345

3. The Variegated Pitpit — — 346
4. The Blue-Capped Pitpit — — 347
5. The Guira-Beraba — — 348

The Yellow Wren — — 350
The Great Yellow Wren — — 356
The Common Wren — — 357
The Gold-Crefted Wren — — 366
VARIETIES *of the Gold-Crefted Wren* — 374
1. The Ruby-Crowned Wren — ib.

2. The

CONTENTS.

Page
2. The Red-Headed Wren — 375
3. The Scarlet-Crowned Blue Titmouſe 376

The Titmouſe-Wren — — 377
The Titmice — — 379
The Great Titmouſe, or Ox-Eye — 394
The Colemouſe — — 401
VARIETIES *of the Colemouſe* — 404
1. The Marſh Titmouſe, or Blackcap — ib.
2. The Canada-Titmouſe — — 407
3. The White-Throat — — 408
4. The Creeper — — — 410

The Blue Titmouſe — — 412
The Bearded Titmouſe — — 416
The Penduline Titmouſe — 420
The Languedoc Titmouſe — — 429
The Long-Tailed Titmouſe — 432
The Cape Titmouſe — — 439
The Siberian Titmouſe — — 441
The Creſted Titmouſe — — 443
FOREIGN BIRDS *which are related to the Titmice* 447
1. The Creſted Titmouſe of Carolina — ib.
2. The Collared Titmouſe — 449
3. The Yellow-Rump Titmouſe — 450
4. The Yellow-Throated Gray Titmouſe 451
5. The Great Blue Titmouſe — 452
6. The Amorous Titmouſe — 454
The Black Titmouſe — — 456

The Nuthatch — — — 458
VARIETIES *of the Nuthatch* — 465
1. The Little Nuthatch — 466
2. The Canada Nuthatch — ib.
3. The Black-Creſted Nuthatch — 467

CONTENTS.

Page

4. The Little Black-Crested Nuthatch — 468
5. The Black-Headed Nuthatch — ib.
6. The Little Brown-Headed Nuthatch 469

FOREIGN BIRDS *related to the Nuthatch* 471
1. The Great Hook-Billed Nuthatch — ib.
2. The Spotted Nuthatch — 472

The Creepers — — 473
The Common Creeper — — 476
VARIETY *of the Creeper* — 480
The Great Creeper — — ib.

The Wall Creeper — — 481
FOREIGN BIRDS *of the ancient Continent, which are related to the Creepers* — 485
1. The Soui-Manga — — 487
Variety of the Soui-Manga — 488

2. The Red-Breasted Purple-Chesnut Soui-Manga 489
Varieties of the preceding — 490
1. The Little Creeper, or Soui-Manga ib.
2. The Creeper, or Soui-Manga 491

3. The Red-Breasted Violet Soui-Manga 492
4. The Purple Soui-Manga — 493
5. The Collared Soui-Manga — 494
6. The Purple-Breasted Olive Soui-Manga 498
7. The Angala Dian — — 502
8. The Iris Soui-Manga — 504
9. The Red-Breasted Green Soui-Manga 505
10. The Black, White, and Red Indian Creeper, or Soui-Manga — 506
11. The Bourbon Soui-Manga — 507

The

CONTENTS.

Page

The Long-Tailed Soui-Mangas — 508

1. The Long-Tailed Violet-Hooded Soui-Manga 509
2. The Long-Tailed Soui-Manga of a Gloffy Gold Green — — 511
3. The Great Green Long-Tailed Soui-Manga 512
4. The Creeper-Billed Red Bird — 514
5. The Creeper-Billed Brown Bird — 516
6. The Creeper-Billed Purple Bird — 518

The American Guit-Guits — 519

1. The Black and Blue Guit-Guit — 520
   Variety of the Black and Blue Guit-Guit 522
2. The Black-Headed Green and Blue Guit-Guit 524
   Varieties of the Black-Headed Green and Blue Guit-Guit — — 525
   1. The Black-Headed Green Guit-Guit ib.
   2. The White-Throated Green and Blue Guit-Guit — — 526
   3. The All-Green Guit-Guit 527
3. The Spotted Green Guit-Guit — ib.
4. The Variegated Guit-Guit — 529
5. The Black and Violet Guit-Guit — 530
6. The Sugar-Bird — — 532

No. 115

THE SKY-LARK.

THE

SKY-LARK.

L'Alouette, *Buff.*
Alauda Arvensis, *Linn. Scop. Brun. Mull. Kram. &c.*
Alauda Vulgaris, *Ray. Will. & Briss.*
Alauda Cælipeta, *Klein.*
The Sky-lark or Field-lark, *Lath. Penn. Alb. Will.**

THIS bird, which is now widely diffused, seems to have been a more ancient inhabitant of Gaul than of Italy; for its Latin

* The ancient Greek name, Κορυδος, or Κορυδαλος, is derived from κορυς, a helmet, on account of its crest: it had the epithets, χαμαιζηλος, ωδικος, αγλαιος, and ευπτερος; i. e. *attached to the ground, excelling in song, distinguished in plumage, and of vigorous wing.*

The Latin name, *Alauda*, is, according to Pliny, Suetonius and Varro, of Gaulish extraction. Hence the present French term *Alouette.*

In Italian, it is called *Lodola, Petronella, Allodola, Alodetta.*

In Spanish, *Eugniada.*

In old Saxon, *Leefwerc* or *Leeurich.*

In modern German, *Heid-lerche, Sang-lerche, Himmel-lerche, Korn-lerche, Grosse-lerche, Field-lerche,* &c. i. e. *Heath-lark, Song-lark, Sky-lark, Corn-lark, Great-lark,* and *Field-lark, &c*

In Dutch, *Leeurich*:—in Sweden, *Laerka.*

In old and provincial English, *Wild-lark, Heath-lark,* and *Laverock.*

name, *Alauda*, is, according to the beſt informed Roman authors, of Gauliſh origin*.

The Greeks were acquainted with two ſpecies of larks: the one wore a tuft on its head, and for that reaſon termed Κορυδος or Κορυδαλος, which the Latins render *Galerita* or *Caſſita*; the other, which wanted the tuft, is the ſubject of this article †. Willughby is the only author that I know, who mentions that the latter ſometimes briſtles the feathers on its head ſo as to form an occaſional creſt, and I have myſelf aſcertained this fact in regard to the male; and thus it is alſo entitled to the epithet *Galerita*. The Germans call it *Lerche*, which in many provinces is pronounced *leriche*, and is obviouſly intended to imitate its notes ‡. The Honourable Daines Barrington reckons it among

* The Celtic name is *Alaud*; whence *Aloue*, and afterwards *Alouette*, the preſent French name. Probably the ſoldiers of the legion ſtyled *Alauda*, wore on their helmet a tuft ſomething like that of the ſky-lark. Schwenckfeld and Klein, who ſeem to have never read Pliny, derive *Alauda* from *laus*, *a laude*; becauſe, according to the former, this bird riſes ſeven times a-day to ſing praiſes to God. It is admitted that all creatures atteſt the exiſtence and glory of their Maker: but to ſuppoſe the ſmall birds have ſtated hours of devotion, and to ground this inference on the accidental reſemblance of words in two different languages, is a very puerile idea.

† Ariſtotle, *Hiſt. Anim.* lib. ix. 25.

‡ "It prolongs its *tirile, tirile, its tirile*." Linnæus, *Syſtema*.

the

the beſt of the ſinging * larks; and as it copies the warble of every other bird †, he terms it a mocking bird: but if it lays no claim to originality of muſic, the delicacy and flexibility of its organs of voice ſmooth and embelliſh whatever it imitates.

In the ſtate of freedom, it commences its ſong early in the ſpring, which is its ſeaſon of love, and continues to warble during the whole of the ſummer. It is heard moſt in the morning and evening, and leaſt in the middle of the day ‡. It is one of thoſe few birds which chant on wing: the higher it mounts, the more it ſtrains its voice; and when it ſoars beyond the range of our ſight, its muſic ſtill diſtinctly ſtrikes our ear. Muſt we impute this ſwell to the joyous elevation of its ſpirits, or the throbbing emotions of love; or muſt we regard it as a ſort of call, the ſignal of common danger? The rapacious tribes, truſting to their ſtrength, and meditating deeds of carnage, proceed with

* "Its ſong is delightful for its variety; is full of ſwells and falls." *Olina.*

† Friſch.—Schwenckfeld pretends that it ſings better than the creſted lark: others prefer the warble of the latter; Kœmpfer that of the Japaneſe lark, which is perhaps not of the ſame ſpecies. *See* particularly Barrington's paper in the Philoſophical Tranſactions for 1773, vol. lxiii. part 2.

‡ Aldrovandus. This may be the caſe in the hot climates of Italy and Greece; but in our temperate climates, the ſky-lark is not obſerved to pauſe at noon.

cautious and dark ſilence; the little harmleſs birds have nothing to depend on but their numbers; and their clamorous notes may ſummon the ſtragglers together, and at leaſt inſpire a pleaſing, though often a vain confidence.—The ſky-lark ſeldom ſings on the ground; where however it conſtantly remains, except when it flies, for it never perches on trees. It may be reckoned among the pulverent birds*; and if it be kept in the cage, we muſt be careful to lay a bed of ſand in a corner, that it may welter at its eaſe, and procure ſome relief of the vermin which torment it.

It has been fabled, that theſe birds have an antipathy to certain conſtellations; to *Arcturus*, for inſtance; and that they were ſilent when that ſtar riſes heliacaly†. This ſeems to mark the time of moulting.

I need not ſtop to deſcribe a bird ſo well known: I ſhall only obſerve, that the principal characters are theſe: the middle toe is cloſely connected, by the firſt phalanx, to the outermoſt on each foot; the nail of the hind toe is very long and almoſt ſtraight, the anterior nails very ſhort and ſlightly curved; the bill not weak, though awl-ſhaped; the tongue broad, hard and forked; the noſtrils round, and half bare; the ſtomach fleſhy, and large in proportion to the ſize of the bird; the liver di-

* Ariſtotle, *Hiſt. Anim.* lib. ix. 49.
† Anton. Miraldus *apud Aldrovandum.*

vided

vided into two very unequal lobes, the left one apparently checked in its growth by the preſſure of the ſtomach; the inteſtinal tube is nine inches long, and two very ſmall *cœca* communicate with it; there is a gall-bladder. The plumage is of a dingy caſt; the tail contains twelve quills, and the wings eighteen, of which the middle-ſized ones are cut almoſt ſquare and notched, a character common to all the larks*. I ſhall add, that the males are rather browner than the females†; that they have a black collar, and that they have more white on the tail‡; that they have a bolder aſpect, and are rather larger, though they never weigh more than two ounces; and laſtly, that, as in almoſt all other ſpecies, they excluſively poſſeſs the talent of ſong. Olina ſeems to ſuppoſe that their hind-nail is longer §; but I ſuſpect, with Klein, that this depends as much upon the age as upon the ſex.

In the opening of the vernal ſeaſon, the male feels the ardour of love; he mounts into the

* Briſſon and Willughby.

† Friſch and Aldrovandus. I believe that the larks of Beauce, which are ſold at Paris, are browner than thoſe of Burgundy. Some individuals are more or leſs of a ruſt colour, and have more or leſs of the wing-quills edged with that colour.

‡ Albin.

§ Geſner affirms, that he ſaw one of theſe nails about two inches long; but he does not tell us whether the bird was a cock or a hen.

air, warbling his impaſſioned ſtrains; and ranges over an extent proportioned to the number of females, till he deſcries his favourite, and inſtantly he darts to the ground, and conſummates the union. The impregnated female ſoon proceeds to form her neſt; ſhe places it between two clods of earth, and lines it with herbs and dry roots *, and is equally careful in concealing as in building it: accordingly few neſts of ſky-larks are found, in compariſon with the number of theſe birds †. Each female lays four or five eggs, which are greyiſh, with brown ſpots; ſhe ſits only fifteen days at moſt, and employs ſtill leſs time in training and educating her young. This expedition has often deceived perſons who intended to rob the neſtlings, and Aldrovandus among the reſt ‡. The ſame circumſtance renders probable what Aldrovandus and Olina aſſert, that ſhe has three hatches in the year; the firſt, in the beginning of May; the ſecond, in the month of July; and the laſt, in the month of Auguſt: but if this take place, it muſt be in the warm countries, where incubation is more forward, where the young are ſooner emancipated, and where the mother is ſoon in a condition to renew her loves. In fact, Aldrovan-

* Fowlers ſay, that the lark's neſt is better conſtructed than that of partridges and quails.

† *Deſcription of* 300 *Animals.*

‡ Tom. ii. p. 834.

dus

dus and Olina wrote in the climate of Italy; but Frifch, whofe obfervations apply to Germany, mentions only two hatches annually; and Schwenckfeld takes notice of one only in Silefia.

The young keep at a little diftance from each other; for the mother does not always gather them under her wings. She flutters over their heads, watches them with a truly maternal affection, directing their motions, anticipating their wants, and guarding them from danger.

The inftinctive warmth of attachment which the female fky-lark bears to her young, often difcovers itfelf at a very early period; and even before fhe is capable of difcharging the functions of a mother, which might be fuppofed to precede, in the order of nature, the maternal folicitude. A young hen-bird was brought to me in the month of May, which was not able to feed without affiftance; I caufed it to be educated, and it was hardly fledged when I received, from another place, a neft of three or four callow fky-larks. She took a ftrong liking to thefe new-comers, which were fcarcely younger than herfelf; fhe tended them night and day, cherifhed them beneath her wings, and fed them with her bill. Nothing could divert her tender offices: if the young were torn from her, fhe flew back to them as foon as fhe was liberated, and would not think of effecting her own escape, which she might have done an hundred times. Her affection grew upon her; she neglected food and drink; she now required the same support as her adopted offspring, and expired at last, consumed with maternal anxiety. None of the young ones survived her; they died one after another: so essential were her cares, which were equally tender and judicious.

The most common food of the young sky-larks is worms, caterpillars, ants eggs, and even grasshoppers; which has justly procured them much regard in countries subject to the ravages of these destructive insects*. After they are grown up, they live chiefly on seeds, herbage, and in short, on all vegetable substances.

It is said, that those destined for song, should be caught in October or November, the males being preferred as much as possible†; and when they are furious and untractable, they must be pinioned, lest they dart with too great violence against the roof of the cage, and break their skull. They are easily tamed, and become so familiar that they will eat off the table, and even alight on the hand; but they cannot cling by the toes, on account of the form of the hind toe, which is too long and straight. This is undoubtedly the reason why they never perch on

* Plutarch *de Iside*.

† Albin.

on trees.—It is eaſy to infer, that there ought to be no bars laid acroſs their cage.

In Flanders, the young ones are fed with moiſtened poppy-ſeeds; and after they can eat without aſſiſtance, they are preſented with crumbs of bread, likewiſe ſoaked: but when they begin to ſing, they are given ſheeps' and calves' hearts haſhed with hard eggs *; and to this are added, wheat, ſpelt, and oats previouſly cleaned, millet, linſeed, and the ſeeds of poppy and hemp, the whole being ſteeped in milk †. Friſch tells us, that when they ſubſiſt on bruiſed hemp-ſeed alone, their plumage is apt to turn black. It is alſo ſaid, that muſtard-ſeed is improper food for them; but except this, they may be fed with every other ſort of farinaceous ſeed, and even every thing uſed at our tables, and become in ſome degree domeſtic birds. According to Friſch, they have a ſingular inſtinct of taſting with their tongue, before they venture to ſwallow.—They may be taught to ſing, and to heighten their native warble with all the embelliſhments which our muſic can beſtow. Some cock-larks, after hearing a tune whiſtled with the pipe, have caught the whole, and repeated it more agreeably than any linnet or canary.—Thoſe which remain in the wild ſtate, inhabit, during the

* Albin.

† Olina—*Deſcription of* 300 *Animals*—Friſch.

the ſummer, the higheſt and drieſt ſituations; and in winter, they deſcend into the plains, and aſſemble in numerous flocks. In that ſeaſon they are very fat, for then they are almoſt always on the ground, and perpetually feeding. In ſummer, on the contrary, they are very lean: then they always go in pairs, eat ſparingly, ſing inceſſantly, and never alight but to hold the dalliance of love. During ſevere weather, particularly when much ſnow has fallen, they reſort to the margins of the ſprings, where the froſt does not prevail; in ſuch ſeaſons, they crop the graſs, and are even obliged to ſeek their food among the horſe-dung which is dropt on the high roads: yet notwithſtanding, they are ſtill fatter than in any part of the ſummer.

They mount in the air almoſt perpendicularly, and by ſucceſſive ſprings, and hover at a vaſt height. They deſcend to the ground, on the contrary, by an oblique ſweep, unleſs they are threatened by a ravenous bird, or attracted by a beloved mate, and in theſe caſes they drop like a ſtone *.

It is eaſy to conceive, that theſe ſmall birds, which ſoar ſo lofty, may be carried ſometimes far to ſea by a guſt of wind, or even be wafted acroſs the ocean. "As ſoon as we approach the European coaſts, ſays Father Dutertre, we be-

* Olina.

gin

gin to ſee birds of prey, larks, and goldfinches, which are driven off from the land, and are glad to reſt on the maſts and cordage of the veſſels *." Hence Sir Hans Sloane ſaw them forty miles at ſea, and the Count Marſigli met with them on the Mediterranean. It is even probable, that thoſe which are found in Pennſylvania, Virginia, and other parts of America, have been driven thither the ſame way. The Chevalier des Mazis informs me, that the larks paſs the iſland of Malta in the month of November; and though he does not ſpecify the particular kinds, the ſky-lark is probably one of them: for Lottinger has obſerved, that a conſiderable flight of ſky-larks into Lorraine, ceaſes exactly at this time; and that as thoſe bred in the country join the train of their viſitors, few ſtay behind; but that ſhortly after, the uſual numbers again appear, whether that others ſucceed to their place, or that they return back, which is the more probable ſuppoſition. However, it is certain that they do not entirely migrate, ſince they are found at all ſeaſons in our province; and conſiderable numbers of them are caught in Beauce, Picardy, and many other parts, during winter. It is indeed the general opinion, in theſe places, that they are not birds of paſſage. If they be abſent for a few days during the exceſſive cold weather, eſpecially

* *Hiſt. des Antilles*, t. ii. p. 55.

after

after long continued ſnow, it is often becauſe they retire under ſome rock or ſome ſheltered cave *, as I have already ſaid they haunt the perennial ſprings. Frequently it happens that they ſuddenly diſappear in the ſpring, when the mild gleams which drew them from their retreats, are ſucceeded by froſts or ſtorms, that drive them back. This temporary concealment of the lark was known to Ariſtotle †, and Klein aſſures us that he aſcertained the fact from his own experience ‡.

This bird is found in all the inhabited parts of both continents, and as far as the Cape of Good Hope, according to Kolben §. It could even ſubſiſt amidſt uncultivated tracts over-

* In the part of Bugey ſituated at the foot of the mountains between the Rhône and the Dain, an innumerable multitude of ſky-larks are often ſeen about the end of October, or the beginning of November, for the ſpace of a fortnight. During the intenſe cold which prevailed the laſt fortnight of January 1776, there appeared in the neighbourhood of Pont-de-Beauvoiſin, ſuch prodigious quantities of ſky-larks, that one perſon with a pole killed as many as would load two mules. They took ſhelter in the houſes, and were exceſſively lean. It is evident from theſe two caſes, that the larks did not quit their ordinary reſidence for want of food; but ſtill we cannot abſolutely infer that they are not birds of paſſage. Thevenot ſays, that the larks appear in Egypt in the month of September, and continue there till the end of the year. *Voyage du Levant*, t. i. p. 493.

† *Hiſt. Anim.* lib. viii. 16. "The ſtork hides itſelf, and the blackbird, and the turtle, and the lark."

‡ P. 181.

§ Hiſt. Gen. des Voyages, t. iv. p. 243.

ſpread with heath and junipers; for it is exceedingly fond of theſe ſhrubs*, which ſhelter itſelf and its infant brood from the attacks of the birds of prey. If the ſky-lark accommodates itſelf with ſuch facility to every ſituation and every climate, it ſeems rather ſingular that the bird is not found in the Gold-coaſt, according to Villault, nor even in Andaluſia, if we credit Averroes.

Every perſon knows the various contrivances for catching larks; the nooſe, the trammel, the ſpringe, the draw-net, &c; but the engine moſt commonly employed, is what is called the *lark-net.* A cool morning gladdened by a bright ſun, is choſen for the ſport; a mirror that turns freely on its pivots is provided, and a pair of ſky-larks are uſed as calls; for it is impoſſible to imitate their ſong ſo cloſely as to deceive them, and hence they liſten to no artificial ſubſtitute. But they ſeem the moſt attracted by the mirror; not indeed to admire their image, as ſome have ſuppoſed, on account of the inſtinct which they have in common with almoſt all the other birds which are capable of being tamed, viz. that of ſinging with redoubled vivacity and emulation before a glaſs: their curioſity is raiſed by the dazzling glare which beams from every quarter as the mirror turns round; they perhaps miſtake it for the undu-

* Turner and Longolius *in Geſner*.

lating

lating ſurface of the cryſtal fountain, which in that ſeaſon has to them its peculiar charms. Accordingly great numbers are caught every winter near the tepid ſprings, which they haunt. But the moſt ſucceſsful ſport is that with lime-twigs, as practiſed in French Lorraine*, and in other parts; and which, becauſe it is little known, I ſhall particularly deſcribe. For this purpoſe, 1500 or 2000 willow-rods of about three feet ten inches long are provided, very ſtraight, or at leaſt well ſmoothed; theſe are ſharpened at one of the ends, and even ſlightly burnt, and the ſpace of about a foot from the other end is ſpread with bird-lime. The ſtakes are planted in parallel rows, in a proper ſituation, commonly in fallow-ground, that is likely to harbour a ſufficient number of larks to defray the expence, which is conſiderable. The rows ought to be as wide as to admit a perſon to paſs eaſily between them; and each ſtake ſhould be a foot diſtant from the next one, and placed oppoſite to the interval in the adjacent row. The art conſiſts in planting theſe with great regularity, and quite perpendicular, and ſo that they ſhall retain their poſition as long as they are not touched, but yet ſo ticklish that they ſhall fall the inſtant a lark bruſhes againſt them in its flight.

* M. de Sonini has long practiſed this mode of fowling in his eſtate of Manoncour in Lorraine. The late king Staniſlaus was fond of it, and often honoured it with his preſence.

After

After all these limed-rods are planted, an oblong square is traced, which presents one of its sides to the ground where the larks are lodged; and at each corner a flag is fixed, which serves the fowlers as a mark, and sometimes as a signal for their manœuvres. The number of persons employed must be proportioned to the extent of the field.—About four or five o'clock in the afternoon, according as the autumn is more or less advanced, the company divides into two equal detachments, each conducted by an intelligent leader, who is likewise subject to the direction of the commander in chief, whose station is in the centre. The one detachment assembles at the flag on the right, the other at the one on the left; and each observing the most profound silence, extends itself in an arch, so that they meet at the distance of half a league from the front, and then form one rank, which gradually closes as it advances to the rods, and continually drives the larks before it. About sunset, the middle of the line ought not to be farther than two or three hundred paces from the front; and this is the time when they *charge* *, that is, they proceed cautiously, stop, or lie flat on the ground, rise up or push forward, according to the commands of their leader: and if all these manœuvres are well executed and properly directed, the greatest part of the larks

* In French, *Donne*.

inclosed

inclosed by the troop, and which now mount no higher than three or four feet, rush forward, and are caught among the limed rods; and falling to the ground with these, they can be picked up by the hand. If it is not too late, a second line is made on the opposite side, at the distance of fifty paces, which drives back the larks that had escaped; and is called *tacking about* *. The idle spectators are detained a little behind the flags, to avoid confusion.

An hundred dozen of larks or more are sometimes caught in one of these sweeps, and it is reckoned very bad sport when only twenty-five dozen are got. Coveys of partridges, and even owls, are also led sometimes into the snare; but such accidents are regarded as vexatious, since they scare away the larks. A hare likewise which happens to cross the field, or any sudden or uncommon noise, spoils the sport.

Many of the sky-larks are also destroyed in summer by the voracious tribes; for they are the ordinary prey of even the smallest of these: and the cuckoo, which has no nest of its own, frequently substitutes its eggs in the place of theirs †. Yet notwithstanding the havock which is made among them, they are extremely nu-

* In French, *Revirer*.

† "The cuckoo breeds in strangers' nests, especially in those of the ring-pigeon, of the petty-chaps, and of the ground-lark." *Hist. Nat. Anim.* lib. ix. 24.

merous;

merous; which proves their great fecundity, and adds to the probability of the aſſertion, that they have three hatches annually. For ſo ſmall an animal, the ſky-lark is long-lived; the term being ten years, according to Olina; twelve, according to others; twenty-two, according to the account of a perſon of veracity; and even twenty-four, if we believe Rzacynſki.

The ancients pretended that the fleſh of the lark, boiled, roaſted, or even burnt and reduced to aſhes, was a ſort of ſpecific in the colic. On the contrary, ſome modern obſervations ſhew that it often occaſions that diſorder, and Linnæus judges it improper food for perſons afflicted with gravelly complaints. The moſt probable account is, that this meat is very wholeſome and pleaſant when fat; and that the pains in the ſtomach, or gripes in the bowels, which are ſometimes felt after eating, are owing to ſome portions of their ſmall bones that have been inadvertently ſwallowed, and which are very minute and very ſharp. The weight of the bird varies, according to the quantity of the fat, from ſeven or eight gros to ten or twelve.

Total length, about ſeven inches; the bill, ſix or ſeven lines; the hind nail ſtraight, and meaſures ten lines; the alar extent, twelve or thirteen inches: the tail, two inches and three quarters, a little forked, conſiſting of twelve quills, and exceeding the wings by twelve lines.

 (A) Spe-

(A) Specific character of the Field or Sky Lark, *Alauda Arvensis*: "The two outermost quills of its tail are white lengthwise externally, the intermediate ones are ferruginous on the inside." Mr. Pennant tells us, that in the neighbourhood of Dunstable four thousand dozens of larks are usually caught for the London market, between the fourteenth of September and the twenty-fifth of February. In fine weather, the larkers use clap-nets with bits of mirror and a decoy: at other times, they employ a trammel net and a pointer dog.

VARIETIES of the SKY-LARK.

I. THE WHITE SKY-LARK. Brisson and Frisch properly consider this as a variety of the foregoing species. In fact, it is a true sky-lark, which according to Frisch comes from the north, like the white sparrow and stare, the white swallow and petty-chaps, &c. in all which the plumage retains the impression of their natal climate. Klein rejects this opinion, because at Dantzic, which is situated farther north than the countries where white larks sometimes appear, not one has been found in the course of half a century. If I were to decide this point, I should say that the assertion of Frisch,

Friſch, that all the white larks arrive from the boreal tracts, is too general, and that the objection which Klein makes is by no means concluſive. In fact, obſervations evince that white larks are found in other countries beſides thoſe of the north; but it is obvious that from Norway, Sweden, and Denmark, they could more eaſily enter the weſtern part of Germany, which is ſeparated from theſe countries by only a narrow ſea, than croſs the Baltic, and reach the mouth of the Viſtula and the coaſts of Pruſſia.—At any rate, beſides the white larks ſometimes found near Berlin according to Friſch, they are often ſeen in the vicinity of Hildeſheim in Lower Saxony. They are ſeldom of a ſnowy white; the ſubject examined by Briſſon was tinged with yellow; but the bill, the feet, and the nails were entirely white.—At the very time when I was writing this article, a white lark was brought to me, which was ſhot under the walls of the little town where I live: the crown of the head and ſome parts of the body were of the ordinary colour; the reſt of the upper ſurface, including the tail and the wings, was variegated with brown and white; moſt of the plumage and even the quills were edged with white; the under part of the body was white, ſpeckled with brown, eſpecially the fore part, and the right ſide; the lower mandible whiter than the upper, and the feet of a dirty white, variegated with brown. This ſubject ſeemed to form the intermediate ſhade between the common lark and that which is of a pure white.

I have ſince ſeen another lark whoſe plumage was perfectly white, except on the head, where there were ſome traces of a grey but half effaced. It was found in the neighbourhood of Montbard. It is not likely that either this or the other lark came from the northern ſhores of the Baltic.

II. THE BLACK SKY-LARK. I join Briſſon in conſidering this as a variety of the common ſky-lark; whether we are to impute it to the bird's feeding on hemp-ſeed, or to any other cauſe. The ſubject which I have directed to be engraved was of a rufous brown at the riſe of the back, and its feet of a light brown.

Albin, who ſaw and deſcribed this bird from nature, repreſents it entirely of a dull brown and reddiſh, verging to black; except however the back of the head, which is of a dun yellow, and below the belly, where there are ſome feathers edged with white; the feet, the toes, and the nails were of a dirty yellow. The ſubject from which Albin formed his deſcription was caught with a net in a meadow near Highgate; and it appears that there ſuch birds are ſeldom met with.

Mauduit aſſures me that he ſaw a lark which was perfectly black, and had been caught in the plain of Mont-rouge near Paris.

THE

RUFOUS-BACKED LARK.

Alauda Rufa, *Gmel.*
L'Alouette Noire a Dos Fauve, *Buff.*

HAD not this bird been brought by Commerſon from Buenos-Ayres; were it not much ſmaller than our common ſky-lark, and a native of a very different climate, the reſemblance of its plumage is ſo ſtriking that we could not help conſidering it as only a variety of the preceding ſpecies. The head, the bill, the feet, the throat, the fore part of the neck, and all the under ſurface of the body, are of a blackiſh brown; the quills of the wings and of the tail are of a ſomewhat lighter ſhade; the outermoſt of theſe laſt are edged with rufous; the hind part of the tail, the back, and the ſhoulders, are of an orange fulvous; the ſmall and middle coverts of the wings blackiſh, edged alſo with fulvous.

Total length, rather leſs than five inches; the bill, ſix or ſeven lines, the edges of the lower mandible being a little ſcalloped near the tip; the *tarſus*, nine lines; the hind toe, two lines and a half, and its nail four lines, ſlightly bent back; the tail, eighteen lines, ſomewhat forked, conſiſting of twelve quills, and exceeding the wings by ſeven or eight lines. Upon a cloſe view, we perceive that its dimenſions do not differ more than thoſe of the preceding variety.

(A) Specific character of the *Alauda Rufa*: "Its tail quills are brown; the eight middle ones tawny at the edges, the outermoſt white at the edge."

THE

No. 116

THE WOOD-LARK.

THE

WOOD-LARK.

Le Cujelier, *Buff.*
Alauda Arborea, *Linn. Gmel. Scop. Brun. Kram. Briss Will. & Klein.**

I CONCEIVE this bird to differ so much from the common lark as to constitute a distinct species. It is distinguished by its size and its general form, being shorter, rounder, and much smaller, not weighing more than an ounce: by its plumage, whose colours are more dilute, less mixed with white, and its whitish crown is more conspicuous than in the ordinary species: by the dimensions too of its wing quills, the first and outermost being half an inch shorter than the rest. It is discriminated from the sky-lark also by its habits: it perches on trees, though only indeed on the thick branches, because the length of its hind toe, or rather the projection and slight curvature of the nail, will not permit it to cling to the twigs: it haunts

* In Italian, *Tottovilla*: in Danish, *Skow-lerke*: in Norwegian, *Heede-lerke*, *Lyng-lerke*: in German, *Lud-lerche*, *Wald-lerche*.

the uncultivated tracts near copses, or even the verge of young copses; and hence the name of *wood-lark*, though it never penetrates into the woods. Its song too resembles more the warble of the nightingale than that of the sky-lark *; and is heard not only in the day, but, like the former, in the night; both when it flutters on the wing and when it sits on a bough. Heb rt observes, that the fifers of the Swiss guards imitate well its notes; hence I conclude, that this bird is common among the mountains of Switzerland †, as in those of Bugey. Its fecundity is inferior to that of the sky-lark; for, though it also lays four or five eggs, and is not so much destroyed, because smaller and less valued, yet its numbers are not so great ‡. It breeds earlier, since its young are sometimes flown in the middle of March §, whereas the common lark does not hatch before the month of May. It is besides more delicate; for, according to Albin, it is impossible to rear the young taken out of the nest. But this holds only in regard to England, and other similar or colder climates; and Olina positively asserts, that in Italy the young are removed from the

* Olina and Albin.

† I am informed that it actually frequents the highest meadows in Switzerland.

‡ British Zoology.

§ Albin.

nest

neſt and raiſed at firſt like the nightingale *, and afterwards fed upon panic and millet.

In every other property the wood-lark bears a cloſe analogy to the ſky-lark. It mounts high, warbling its notes, and hovers in the air: it flies in flocks during the winter colds: it builds its neſt on the ground, and conceals it under a turf: it lives ten or twelve years: it feeds on beetles, caterpillars, and ſeeds: its tongue is forked; its ſtomach muſcular and fleſhy: and it has no craw, but a moderate dilatation of the lower part of the *œſophagus*: its *cæca* are very ſmall †.

Olina remarks that, in the male, the crown of the head is darker than in the female, and its hind nail longer. He might have alſo added, that its breaſt is more ſpotted, and its great wing quills edged with olive, while, in the female, they are bordered with gray. He ſubjoins, that the wood-lark is caught in the ſame manner as the ſky-lark, which is true; but he pretends that this ſpecies is hardly known out of the Pope's territories, which is juſtly controverted by the beſt informed modern naturaliſts. In fact, it would ſeem that the wood-lark is not confined to any one country:

* Willughby remarks, that the ſong of the wood-lark reſembles that of the blackbird.

† Willughby.

it

it is found in Sweden * and Italy †, and is probably ſpread through the intervening countries, and conſequently ſcattered over the greateſt part of Europe ‡.—The wood-lark is pretty fat in autumn, and is then excellent meat.

Albin ſays, that there are three ſeaſons for catching the wood-larks. The firſt is in the ſummer, when the ſmall branchers begin to chirp, before they undergo the moulting.—The ſecond is in the month of September, when they fly in flocks, and roam from one country to another, roving over the paſture grounds, and perching on trees near lime-kilns §. At this time the young birds change their plumage, and are no longer to be diſtinguiſhed from the old ones.—The third ſeaſon, and the moſt favourable for catching the wood-larks, begins with the month of January ‖, and laſts till the end of February, when they ſeparate to pair. The young birds which are then caught, make

* Linnæus. † Olina. ‡ Linnæus. § Kramer.

‖ M. Hebert killed theſe birds during winter in Brie, in Picardy, and in Burgundy: he remarked that, during this ſeaſon, they are found on the ground in the plains, that they are pretty common in Bugey, and ſtill more ſo in Burgundy. On the other hand, M. Lottinger aſſerts that they arrive about the end of February, and retire in the beginning of October. But theſe oppoſite accounts might be reconciled, if of theſe larks, as of the common ſort, ſome are migratory and others ſtationary.

generally

generally the beſt ſingers: they chirp a few days after, and with a clearer tone than thoſe caught at any other ſeaſon.

Total length, ſix inches; the bill, ſeven lines; the alar extent, nine inches (ten, according to Lottinger); the tail, two inches and a quarter, rather forked, conſiſting of twelve quills, and exceeding the wing by thirteen lines.

(A) Specific character of the Wood-lark, *Alauda Arborea*: "Its head is encircled by a white annular belt." Its egg is light gray, with numerous dark and purpliſh dots.

THE

TIT-LARK.

La Farlouſe, ou l'Alouette de Pres, *Buff.*
Alauda Pratenſis, *Linn.* *Gmel.* *Brun.* *Mull.* *Friſ.* *Will.* *Briſſ.*
Alauda Pratorum, *Klein* *.

BELON and Olina mention this as the ſmalleſt of all the larks; but they were unacquainted with the graſhopper-lark, of which we ſhall afterwards ſpeak. The tit-lark weighs ſix or ſeven gros, and its alar extent is only nine inches. The prevailing colour of its upper ſurface is olive, variegated with black on the fore part, and pure olive behind. Its under ſurface is yellowiſh white, with black longitudinal ſpots on the breaſt and the ſides, and the ground colour of the plumage is black. The quills of the wings are almoſt black, edged with olive; and thoſe of the tail are ſimilar, except the outermoſt one, which is edged with white, and the one next it, which is tipt with the ſame colour.

This bird has a ſort of white eyelids, which

* In Italian, *Lodola di Prato*, *Mattolina*, *Petragnola*, *Corriera*: in German, *Brein-vogel*, *Schmel-vogel*.

Linnæus

Nº 117

THE TIT-LARK.

Linnæus adopts as its ſpecific character.—The male has in general more yellow than the female, on the throat, on the breaſt, on the legs, and even on the feet, according to Albin.

The tit-lark is fluſhed at the leaſt noiſe, and ſhoots with a rapid flight; it perches, though with difficulty, on trees; it conſtructs its neſt nearly as the wood-lark, and lays the ſame number of eggs, &c.* But, what diſtinguiſhes it, the firſt quill in its wing is equal to the reſt, and its ſong, though very pleaſant, is leſs varied. Mr. Pennant compares this to a jeering laugh, and Albin, to the warble of the canary finch; but both complain that it is too ſhort and broken. However, Belon and Olina agree that this ſmall bird is eſteemed for the ſweetneſs of its ſong; and I muſt own, that having occaſion to hear it I found it really agreeable, though rather plaintive, and ſimilar to the nightingale's ſtrain, yet not ſo full and connected. It deſerves to be noticed that this one was a female, ſince, in diſſecting it, I diſcovered an *ovarium*: it contained three eggs that were larger than the reſt, which ſeemed to promiſe a ſecond hatch. Olina tells us that the tit-lark is raiſed in the ſame way as the nightingale, but its delicacy renders the ſucceſs extremely precarious; and, as it lives only three or four years, we readily

* Britiſh Zoology.

perceive

perceive the reaſon why the ſpecies is unfrequent, and why the male, when he mounts into the air to deſcry his mate, is obliged to ſweep a much wider circuit than the common lark, or even the wood-lark. Albin indeed pretends that it is long-lived, little ſubject to diſeaſes, and lays five or ſix eggs: but if this were true, the number would be much greater.

According to M. Guys, the tit-lark feeds chiefly upon the worms and inſects for which it ſearches in new-ploughed lands. Willughby actually found beetles and ſmall worms in its ſtomach; and I have myſelf ſeen in it portions of inſects and of ſmall worms, and alſo ſeeds and pebbles. If we believe Albin, it wags its tail from ſide to ſide while it eats.

The tit-larks breed generally in the meadows, and even in low and marſhy grounds *. They make their neſt on the ſurface †, and conceal it artfully. While the female hatches, the male ſits on a neighbouring tree, and riſes at times, ſinging and clapping his wings.

Willughby, who ſeems to have obſerved this bird with great accuracy, ſays properly that its iris is hazel, the tip of its tongue divided into ſeveral filaments, its ſtomach moderately fleſhy, its *cæca* rather longer than in the ſky-lark,

* Britiſh Zoology.

† Belon —Britiſh Zoology.

3 and

and that it has a gall-bladder. All this I have verified; and I ſhall add that there is no craw, and that even the *œſophagus* has ſcarce any inflation at its junction with the ſtomach, and that this ſtomach or gizzard is large in proportion to the body.—I kept one of theſe birds a whole year, and gave it no other food than ſmall ſeeds.

The tit-lark inhabits Italy, France, Germany, England, and Sweden. Albin tells us that it appears (in that part of England, no doubt, where he lived) about the beginning of April with the nightingale, and that it departs about the month of September. It ſometimes begins its flight before the cloſe of Auguſt, according to Lottinger *, and ſeems to perform a long journey. If this be admitted, and if we ſuppoſe that it makes occaſional halts in the temperate countries which it traverſes, it may be among the number of thoſe larks which are ſeen to paſs the iſland of Malta in November. In autumn, that is, in the time of vintage, it haunts the vicinity of the high-roads. M. Guys remarks that it is exceedingly fond of the company of its fellows; and if it cannot obtain the

* M. Lottinger only once ſaw a tit-lark in Lorraine, in the month of February 1774. But, that ſame winter, he ſaw other birds which do not uſually remain in that province, ſuch as green-finches, wagtails, &c. which M. Lottinger aſcribes properly to the mildneſs of that year.

ſociety

ſociety of theſe, it will intermingle with the flocks of finches and linnets which it meets in its paſſage.

Comparing what authors have ſaid of the tit-lark, I perceive differences which diſpoſe one to think that the ſpecies is much ſubject to vary, or that it has been ſometimes confounded with its kindred ſpecies, ſuch as thoſe of the wood-lark and the graſhopper-lark *.

VARIETIES of the TIT-LARK.

THE WHITE TIT-LARK ſcarcely differs from the preceding, except in the plumage, which is almoſt

* The diſpoſition of the ſpots on the plumage is nearly the ſame in the three ſpecies, though the colours of theſe ſpots are different in each, and the habits ſtill more different; but leſs ſo, however than the opinions of authors concerning the properties of the tit-lark. We need only compare Belon, Aldrovandus, Briſſon, Olina, Albin, &c. The colours by which Briſſon characteriſes the ſpecies are not the ſame with thoſe deſcribed by Aldrovandus; who takes no notice of the long hind toe, but ſpeaks of a certain motion of the tail, which the others, except Albin, omit. The latter pretends that the tit-lark is long-lived, and little ſubject to diſeaſes. Olina and Belon, on the contrary, aſſert that it is difficult to be raiſed; Olina poſitively aſſerts that it is ſhort-lived. We need not mention their various opinions reſpecting its ſong.

almoſt univerſally yellowiſh white, but yellower on the wings: its bill and feer are brown. Such was the one which Aldrovandus ſaw in Italy; and though the jeſuit Rzacynſki ranges it among the birds of Poland, I doubt whether it belongs to that country, or at leaſt whether he ever ſaw it, ſince he uſes the words of the Italian naturaliſt without any addition.

Total length, five inches and a half; its bill, ſix lines, the edges of the upper mandible a little ſcalloped near the point; its alar extent, about nine inches; its tail, two inches, ſomewhat forked, compoſed of twelve quills, exceeding the wings eight lines; the hind nail, ſhorter and more hooked than in the preceding ſpecies.

(A) Specific character of the Tit-lark, *Alauda Pratenſis*: "Above it is greeniſh brown, its two outermoſt tail quills white externally, a white line on its eye-brows." Its egg is roundiſh; duſky red, with numerous ſmall ſpots.

FOREIGN BIRD *related to the* TIT-LARK.

THE

LOUISIANA LARK.

La Farlouzzanne, *Buff.*
Alauda Ludoviciana, *Gmel.*

I SAW this bird at M. Mauduit's, and it ſeems to reſemble much the tit-lark. Its throat is of a yellowiſh grey; the neck and breaſt ſtreaked with brown on the ſame ground: the reſt of the under ſurface of its body is fulvous: the upper ſurface of its head and of its body is mixed with greeniſh brown, and with blackiſh; but as theſe colours are dingy, they contraſt little with each other, and by their mixture they form an almoſt uniform dull brown: the ſuperior coverts are greeniſh brown with no addition: the quills of the tail are brown; the outermoſt one conſiſts of blackiſh brown and white, the white being turned outwards, and the next quill tipt with white: the quills and the ſuperior coverts of the wings are of a blackiſh brown, edged with lighter brown.

Total length, near ſeven inches; the bill, ſeven

ſeven lines; the *tarſus*, nine lines; the hind toe with its nail, rather leſs than eight lines, and this nail rather more than four lines, ſlightly curved; the tail, two inches and a half, and exceeds the wings by ſixteen lines.

(A) Specific character of the *Alauda Ludoviciana*: "The laſt quills but one of the tail are tipt with white; the outermoſt are partly brown, partly white."

THE

GRASHOPPER-LARK.

L'Alouette Pipi, *Buff.*
Alauda Trivialis, *Linn.* & *Gmel.*
Alauda Sepiaria, *Briss.*
Piep-Lerche, *Fris.*
Small Lark, *Will.* & *Ray.*
Pippit Lark, *Alb.*
The Grashopper Warbler, *Lath.*

THIS is the smallest of all the French larks. The German epithet *piep* *, and the English *pippit*, allude evidently to its sibilous notes; and such appellations ought always to be preferred, as the most precise and expressive. Its cry, especially in winter, is compared to that of the grashopper; but is rather stronger and shriller. It utters this both when on the wing, and when perched on the tallest branches among the bushes: for though its hind nail be very long (yet not so long or so straight as that

* The German name *Piep-lerche* is formed from the Latin *pipio*, which signifies to utter a feeble shrill cry like chickens. In modern English *pip* expresses the same, and is pronounced *peep* in the northern parts of the kingdom.

Nº 118

THE GRASSHOPPER LARK.

of the ſky-lark), it ſits on the ſmall twigs, claſping with its fore claws. It alſo reſts on the ground, and runs very nimbly.

In the ſpring, when the cock bird ſings on a branch, he performs it with much action: he looks big, diſplays his wings, and gives every mark of ardent emotion. At intervals he riſes to a conſiderable height, hovers ſome ſeconds, and drops almoſt in the ſame place, continuing all the time to ſing; and his tones are ſoft, harmonious, and clear.—This little bird builds its neſt in ſolitary ſpots, and conceals it beneath a turf; and its young often become the prey of the adders. It generally lays five eggs, marked with brown near the large end. Its head is rather long than round; its bill delicate and blackiſh; the edges of its upper mandible ſcalloped near the tip; its noſtrils half covered by a convex membrane of the ſame colour with the bill, and partly concealed under the ſmall feathers which cover it before: there are ſixteen quills in each wing; the upper ſurface of its body is of a greeniſh variegated brown; the under ſurface, of a yellowiſh white, ſpeckled irregularly on the breaſt and neck: the ground colour of its plumage is deep cinereous: laſtly, there are two whitiſh ſtripes on the wings, which Linnæus has made one of the characters of the ſpecies.

The grafhopper-larks appear in England about the middle of September, and great numbers of them are then caught in the environs of London*. They frequent the heaths and plains, and flutter at a moderate height. Some generally remain during the winter in the fens near Sarbourg.

We may infer from the flender form of its bill, that the grafhopper-lark feeds chiefly on infects and fmall feeds; and from its diminutive fize, that it is not long-lived. It is found in Germany, in England, and in Sweden, according to the *Syftema Naturæ* of Linnæus, though he takes no notice of it in his *Fauna Suecica*, at leaft in the firft edition.

It is moderately tall. Total length, about five inches and a half; its bill, fix or feven lines; its hind toe, four lines, and its nail five; its alar extent, eight inches and one-third; its tail two inches, and exceeding the wings an inch †; its inteftinal tube, fix inches and a half; the œfophagus, two inches and a half, dilated before its infertion in the gizzard, which is mufcular; there are two fmall *cæca*; I could find no gall-bladder: the gizzard occupies the left fide of the lower belly, and is covered by the liver, and not by the inteftines.

* Albin.

† Compofed of ten quills, according to a good obferver; but I fufpect that two had been plucked.

(A) Specific

(A) Specific character of the Grashopper-lark, *Alauda Trivialis*: "Its tail quills are brown; the outermost, half white; the second, white at its wedge-like tip, with a double whitish line on the wings." Its egg is light grass green, thinly sprinkled with deeper-coloured specks.

THE

WILLOW-LARK.

La Locuſtelle, *Buff.*
Locuſtella Avicúla, *Will.*

THIS ſpecies is ſtill inferior to the preceding, and indeed it is the ſmalleſt of all the European larks. We are indebted to the *Britiſh Zoology* for the account of it. It annually viſits ſome willow-hedges near a pool in the pariſh of Whiteford, Flintſhire; where it continues the whole ſummer. Its habits and general form are the ſame as thoſe of the graſhopper-lark. With regard to its plumage, the head and the upper ſurface of the body are yellowiſh brown marked with duſky ſpots; the quill feathers are brown, edged with dirty yellow; the tail feathers, deep brown; it has a ſort of whitiſh eye-brows, and the under ſide of the body white tinged with yellow *.

THE

* "Nothing can be more amuſing, ſays Mr. White, than the whiſper of this little bird, which ſeems to be cloſe by though at an hundred yards diſtance; and, when cloſe at your ear, is ſcarce any louder than when a great way off. Had I not been a little acquainted with inſects, and known that

THE

MEADOW-LARK.

La Spipolette, *Buff.*
Alauda Campeſtris, *Linn. Gmel. Brun. Mull. Briſſ. &c.*
Alauda Dumetorum, *Klein* *.

THIS bird is rather larger than the tit-lark. Its hind toe is very long, as that of the ſky-lark, but its body is ſlenderer: it is alſo diſtinguiſhed by the ſhake of its tail, ſimilar to that of the wagtail and of the tit-lark. It inhabits heaths and uncultivated tracts, and frequents the oat-ſtubble ſoon after the corn is reaped, where it gathers in numerous flocks.

that the graſhopper kind is not yet hatched (18th April), I ſhould have hardly believed but that it had been a *locuſta* whiſpering in the buſhes. The country people laugh when you tell them that it is the note of a bird. It is a moſt artful creature, ſculking in the thickeſt part of a buſh; and will ſing at a yard diſtance provided it be concealed. . . . In a morning early, and when undiſturbed, it ſings on the top of a twig, gaping, and ſhivering with its wings." *Natural Hiſtory of Selborne*, p. 45.

* In Italian, *Spipoletta*: in German, *Gickerlin*, *Brachlerche*, *Gereut-lerche*, *Kraut-lerche*: in Daniſh, *Mark-lærke*: in Poliſh, *Zdzbto*.

In ſpring, the male perches to invite or to diſcover his mate; and ſometimes he mounts into the air, ſinging with all his might, and again returns quickly to the ground, which is always the ſcene of their amours.

If a perſon happens to come near the neſt, the female betrays it by her cries; and this evinces a different inſtinct from that of the other larks, which, when danger threatens, remain ſilent and fixed.

Willughby ſaw a neſt of the meadow-lark in a furze-buſh cloſe on the ground; and formed of moſs, lined with ſtraw and horſe-hair.

People have had the curioſity to raiſe the young males for the ſake of their ſong; but it requires attention and care. The cage muſt at firſt be covered with a green cloth, little light muſt be admitted, and a profuſion of ants' eggs muſt be furniſhed. After they are accuſtomed to feed in their priſon, the ſupply of the ants' eggs may be abridged by degrees, and bruiſed hemp-ſeed ſubſtituted, mixed with flour and yolks of eggs.

The meadow-larks are caught with the drag-net, like the ſky-larks, and alſo with lime-twigs, which are placed in the trees which they haunt. They aſſociate with the finches; and it appears even that they arrive and depart along with theſe.

The

The males are hardly to be diſtinguiſhed from the females by their exterior appearance; but if another male be preſented ſhut in a cage, they will inſtantly attack it, as an enemy or a rival *

Willughby ſays, that the meadow-lark differs from all the other larks by the blackneſs of its bill and feet: he adds, that its bill is ſlender, ſtraight, and pointed, and the corners of its mouth edged with yellow; that it has not, like the wood-lark, the firſt quills of the wings ſhorter than the ſucceeding; and that in the male the wings are rather darker than in the female.

This bird is found in Italy, Germany, England, Sweden, &c.† Briſſon conſiders the Jeſſop meadow-lark as the ſame ſpecies with his, though the hind nail is much longer in the latter; but that nail varies much according to age, ſex, &c. There is a wider difference between the meadow-lark of Briſſon and that of Linnæus, though both naturaliſts regard them as the ſame kind: in that of the latter, all the quills of the tail, except the two middle ones, are white from their origin to half their length; but in that of the former, the two outermoſt quills only are white; not to mention many other

* Friſch.

† Aldrovandus and Willughby.

minute

minute differences, which taken together are ſufficient to conſtitute a variety.

The meadow-larks live on ſmall ſeeds and inſects. Their fleſh, when fat, is excellent.—The head and all the upper ſurface of the body are dun grey, tinged with olive; the eye-brows, the throat, and all the under ſurface of the body are yellowiſh white, with brown oblong ſpots on the neck and breaſt; the quills and coverts of the wings, brown, edged with lighter brown; the quills of the tail blackiſh, except the two intermediate ones, which are brown gray, the outermoſt edged with white, and the next one tipt with the ſame: laſtly, the bill is blackiſh, and the feet brown.

Total length, ſix inches and a half; the bill, ſix or ſeven lines; the alar extent, above eleven inches; the tail, two inches and a half, ſomewhat forked, and compoſed of twelve quills; it projects fifteen lines beyond the wings.

(A) Specific character of the Meadow-lark, *Alauda Campeſtris*: "Its tail quills are brown; the lower half, except two intermediate quills, white; the throat and breaſt yellow." Its egg reſembles much that of the ſky-lark, only it is ſcarce one half the ſize, and its tints are lighter.

THE

THE

ITALIAN LARK.

La Girole, *Buff.*
Alauda Italica, *Gmel. & Briss.*
Giarola, *Ray, Will. & Aldrov.*

BRISSON ſuſpects, with much probability, that the ſubject obſerved by Aldrovandus was a young bird, whoſe tail, being extremely ſhort and conſiſting of very narrow feathers, was not entirely formed, and in which the junction of the mandibles was edged with yellow. But I ſhould imagine he ought to have drawn another inference beſides, that it was only a variety of the common ſpecies derived from age, ſince Aldrovandus, the only author who mentions it, ſaw no more than one ſpecimen. Its ſize was the ſame with that of the ſky-lark, and it had the chief character, which is the long nail projecting from each foot. The plumage of the head and of the upper ſide of the body was variegated with cheſnut, with lighter brown, with whitiſh, and with bright rufous. Aldrovandus compares it to that of the quail, or of the woodcock. The under ſurface

ſurface of the body was white; the back of the head encircled with a ſort of whitiſh crown; the quills of the wings brown cheſnut, edged with a lighter colour; thoſe of the tail, at leaſt the four middle ones, of the ſame colour; the pair following, divided by cheſnut and white; and the laſt pair entirely white; the tail ſomewhat forked, and an inch long; the ground colour of the plumage cinereous; the bill red, with a large opening; the corners of the mouth yellow; the feet fleſh-coloured; the nails whitiſh; the hind nail ſix lines in length, almoſt ſtraight, and only a little curved at the tip.

This bird was killed near Bologna, about the end of May. I wiſh thoſe naturaliſts who have an opportunity of obſerving it, would refer it to its true ſpecies; for I much doubt whether it forms a ſeparate ſpecies. Ray ſuppoſed that it belonged to that of the wood-lark, and differing only in the colours of its tail quills; but it is equal to the ſky-lark, and conſequently much larger than the wood-lark; and if with Briſſon we reckon it a young bird, this diſtinction will be the more important.

(A) Specific character of the *Alauda Italica*: "Its middle tail quills are bay, the laſt but one white at the tip, the two outermoſt entirely white."

THE

THE

CALANDRE, OR LARGE LARK.

Alauda Calandra *, *All the Naturaliſts.*

OPPIAN, who lived in the ſecond century of the Chriſtian æra, is the firſt of the ancients who mentions this lark; and he deſcribes the beſt method for catching it, which is preciſely the ſame as that ſince propoſed by Olina, viz. to ſpread a net near the brook whither that bird uſually reſorts to drink.

This bird is larger than the ſky-lark; its bill is ſtronger and ſhorter, ſo that it is able to bruiſe grain; and the ſpecies is leſs numerous, and not ſo widely ſpread. In other reſpects, the calandre reſembles exactly the ordinary lark. Its plumage, its port, its ſhape, its habits, and its tones, are the ſame. Its warble is perhaps more ſonorous than that of the ſky-lark, but is as pleaſant †; for, in Italy, it is an uſual

* Oppian termed it Καλανδρα, and gave it the epithet of Μεγαλοστατος, or *Largeſt.*—Hence all the modern names: In Italian and Spaniſh, *Chalandria:* in German, *Kalander* or *Galander*, &c.

† Belon.

compliment

compliment to ſay one ſings like a calandre *. Like the common ſark, alſo, it can imitate exactly the notes of ſeveral birds, ſuch as the goldfinch, the linnet, the canary, &c. and even the chirping of young chickens, and the love-ſquall of the ſhe-cat †; in ſhort, every ſound adapted to its organs, and impreſſed in its tender age.

To have good ſingers, we muſt, according to Olina, take the young calandres from the neſt, at leaſt before the firſt moult, and preferring thoſe eſpecially which are hatched in Auguſt. We begin with a paſte mixed partly with ſheep's heart, and afterwards add ſeeds and crumbs of bread, taking care always to lay rubbiſh in the cage for whetting their bill, and alſo ſand for them to welter in when teaſed with vermin. But, in ſpite of all our attention, we ſhall derive little pleaſure the firſt year; for the calandre is ſlowly faſhioned into habits of ſlavery. In the beginning, we ſhould even pinion their wings, and inſtead of the top of the cage we ſhould ſubſtitute a ſpread canvaſs ‡. But after they are reconciled to their ſituation, and have acquired the proper bias, they will ſing inceſſantly, grow ſo fond of repeating their own warble, or that of other birds, as ſoon to neglect their food §.

* Aldrovandus. † Olina. ‡ Idem. § Geſner.

The

The male is larger, and blacker round the neck; the female has only a very narrow collar *. Some inftead of a collar have a large black fpace, and fuch was the one that we have figured.—The calandre neftles on the ground like the common lark, under a graffy tuft, and lays four or five eggs. Olina adds, that it lives only four or five years, and confequently is far from gaining the age of the fky-lark. Belon conjectures that it forms flocks like the laft, and fubjoins that it is never feen in France, unlefs it be brought hither: but this affertion relates only to Mans and the contiguous provinces; for the fpecies is frequent in Provence, where it is called *coulaffade* on account of its black collar, and where it is ufually bred for the fake of its fong. With regard to Germany, Poland, Sweden, and the other northern countries, it feems not to vifit them. It is found in Italy, in the Pyrenees, and in Sardinia; and laftly, Dr. Ruffel informed Edwards, that it was common near Aleppo; and Edwards gives a coloured figure

* Edwards.—The perfon who communicated this obfervation to Mr. Edwards, had a method of diftinguifhing the male from the female, in fmall birds. It was to lay them on their back, and blow up their breaft: when the bird was a female, the feathers parted on each fide, leaving the breaft bare. But this method is not certain except in the feafon of hatching. *Gefner*.

of a true calandre, which came, it is ſaid, from Carolina. Itſelf or its progenitors might have been driven acroſs the Atlantic by the fury of the winds, and in that warm climate it would thrive and become naturalized.

Adanſon regards the calandre as intermediate between the ſky-lark and the thruſh: but this analogy muſt be reſtricted to the plumage and the external form; for the habits of the calandre and of the thruſh are very different, eſpecially in the mode of conſtructing their neſts.

Total length, ſeven inches and a quarter; the bill, nine lines; the tail, two inches and one-third, conſiſting of twelve quills, of which the two outer pairs are edged with white, the third pair tipt with the ſame colour, the intermediate pair browniſh gray, and all the reſt black; theſe quills project a few lines beyond the wings: the hind toe meaſures ten lines. [A]

[A] Specific character of the *Alauda-Calandra:* "Its outermoſt tail quill is externally entirely white, the ſecond and third are tipt with white; there is a brown ſtripe on the breaſt."

FOREIGN

FOREIGN BIRDS

ANALOGOUS TO THE CALANDRE-LARK.

I.

THE CAPE-LARK.

La Cravate Jaune, ou Calandre du Cap de Bonne Efperance.
Alauda Capenfis, *Linn. & Briff.*

I DID not fee the individual from which the figure in the *Planches Enluminées* was drawn, but I have examined feveral others of the fame fpecies. In general, the upper fide of the head is brown in the males, variegated with gray; the throat and the top of the neck, beautiful orange; and the collar is edged with black through the whole verge of its circumference: the fame orange forms alfo a fort of eyebrows, and is fcattered in fmall fpots on the little coverts of the wing, and on its anterior border, whofe margin it defines: the breaft is variegated with brown, gray, and blackifh; the belly and loins, with orange rufous; the under furface of the tail, grayifh; the quills of the tail of a brown caft, but the four outer pairs are edged and terminated with white; the quills of the wings brown, and alfo edged, the large ones with yel-

 low,

low, and the middle ones with gray: laſtly, the bill and feet are of a brown gray, variouſly intenſe.

In two females which I obſerved, the collar was not orange, but light rufous; the breaſt ſpeckled with brown on the ſame ground, which became more intenſe as it retired from the fore part: laſtly, the upper ſurface of the body was more variegated, becauſe the feathers were edged with a lighter gray.

Total length, ſeven inches and a half; the bill, ten lines; the alar extent, eleven inches and a half; the hind toe, including the nail, longer than the middle toe; the tail, two inches and a half, ſomewhat forked, conſiſting of twelve quills, and exceeding the wings by fifteen lines.—I ſaw and meaſured a ſpecimen which was an inch longer than the above, and all its other dimenſions were proportionally large. [A]

[A] Specific character of the *Alauda Capenſis*: "Its three lateral tail quills are tipt with white; its throat is yellow margined with black; its eyebrows are yellow."

II. THE

II.

THE SHORE-LARK.

Le Hauſſe-Col Noir, ou l'Alouette de Virginie, *Buff.*
Alauda Alpeſtris, *Linn. & Gmel.*
Alauda Virginiana, *Briſſ.*
Alauda Gutture Flavo, *Cateſby.*

I RANGE this American lark beſide the preceding, to which it is much akin: it differs however by its climate, by its magnitude, and by ſome accidents of its plumage. It ſometimes enters Germany in the time of ſnow; for which reaſon, Friſch terms it *the winter lark.* But we muſt not confound it with the *lulu,* which, according to Geſner, might bear the ſame appellation, ſince it appears when the ground is buried with ſnow. Friſch ſays that it is little known in Germany, and that the place of its retreat is not aſcertained.

Theſe ſhore-larks are alſo caught ſometimes in the neighbourhood of Dantzic, with other birds, in the months of April and December, and one of them lived ſeveral months in a cage. Klein preſumes that they had been driven by a guſt of wind from North America into Norway, or the countries ſtill nearer the pole, whence they would eaſily paſs into milder climates.

It appears too that they are birds of paſſage:

 for

for we learn from Catesby, that they are seen in Virginia and Carolina only in winter, advancing from the north in great flocks; and that in the spring they return by the same route. During their stay they frequent the downs, and feed upon the oat which grows among the sand *.

This lark is of the bulk of the ordinary sort, and its song is nearly the same. The upper side of its body is brown; its bill black; its eyes placed on a yellow bar that rises from the base of its bill; its throat and the rest of its neck of the same colour; and this yellow is partly terminated on each side by a black bar, which, rising from the corners of the mouth, passes under the eyes, and reaches the middle of the neck; it is terminated below the neck by a sort of collar or black gorget: the breast and all the under side of the body are of a deep straw colour.

Total length, six inches and a half; the bill, seven lines; the hind toe and nail still longer than in our lark; the tail, two inches and a half, a little forked, composed of twelve quills, and exceeding the wings ten or twelve lines. [A]

[A] Specific character of the *Alauda Alpestris*: "Its tail quills are white the half of their inside; its throat yellow; the stripe under its eyes, and on its breast, black."

* *Uniola Panicula*, Linn.

III. THE

III.

THE BROWN-CHEEKED PENNSYLVANIAN LARK.

Alauda Rubra, *Gmel.*
Alauda Pennſylvanica, *Briſſ.*
The Red Lark, *Penn.* & *Lath.*

THIS is a migratory lark common to both continents: for Bartram, who ſent the ſpecimen to Edwards, wrote, that it appears in Pennſylvania in the month of March, and is not ſeen after the end of May, but advances northward; and, on the other hand, Edwards found it in the vicinity of London.

This bird is of the ſize of the meadow-lark: its bill is thin, pointed, and of a deep brown colour; its eyes brown, edged with a lighter tint, and ſurrounded by an oval brown ſpot, which deſcends on the cheeks, and bounded by a zone, which is partly white and partly bright fulvous. All the upper ſide of its body is dull brown, except the two outer quills of its tail, which are white; its neck, its breaſt, and all the under ſide of its body, of a reddiſh fulvous ſpeckled with brown: its feet and nails are deep brown, like its bill: its hind nail is very long, but not quite ſo long as in the ſky-lark. A peculiarity of this ſpecies is, that when the wing is cloſed, the third quill reckoning from the body reaches the end of the longeſt quills; which, according to Edwards, is an invariable character of the wagtails. But this is not the only point of analogy between the larks and the wagtails; for we have before ſeen that the meadow and tit-larks have a ſimilar ſhake in their tails. [A]

[A] Specific character of the *Alauda Rubra*: "It is brown, the ſpace about its eyes black, its two outermoſt tail quills white."

THE

THE

MARSH-LARK.

La Rouſſeline, ou l'Alouette de Marais, *Buff.*
Alauda Moſellana, *Gmel.**

THIS lark, which is found in Alſace, is of a middle ſize between the ſky-lark and the tit-lark. The upper part of its head and body is varied with rufous and brown; the ſides of the head, ruſty, marked with three brown ſtripes that are almoſt parallel, and of which the higheſt paſſes below the eye: the throat is of a very light rufous; the breaſt of a deeper rufous, and ſprinkled with little brown ſpots; the belly, and the lower coverts of the tail, light rufous; the quills of the tail and wings blackiſh, and edged with the ſame rufous; the bill and feet yellowiſh.

Like many other ſpecies of this kind, the marſh-lark begins its ſong at dawn, which, according to Rzacynſki, is exceedingly pleaſant. Its name ſhews that it haunts wet ſituations: it often frequents the ſandy margin of the Moſelle, and ſometimes breeds on its banks,

* Rzacynſki terms it the Pine-lark; in Poliſh, *Skowronek Borowy.*

near

near Metz, where it appears annually in October; at which time a few are caught.

Mauduit told me of a rufous lark, in which the feathers on the upper part of the body were tipt with white, and alſo the lateral quills of the tail: this is probably a variety of the marſh-lark.

Total length, ſix inches and a quarter; the bill, eight lines; the *tarſus*, an inch; the hind toe, three lines and a half, ſomewhat curved; the tail, two inches and a quarter, and exceeds the wings by eighteen lines. [A]

[A] Specific character of the Marſh-lark, *Alauda Moſellana*: "It is rufous, below rufous white; its cheeks and breaſt marked with brown lines; its tail black, with a rufous margin."

THE

THE

SIBERIAN LARK.

La Ceinture de Prêtre, ou l'Alouette de Sibérie, *Buff.*
Alauda Flava, *Gmel.*

OF all the birds denominated larks, this is the moſt conſpicuous for beauty of plumage. Its throat, its face, and the ſides of its head, are of a pleaſant yellow, which is ſet off by a black ſpot between the eye and the bill, that joins to another larger one immediately below the eye: its breaſt is ornamented with a broad black girdle: the reſt of the under ſurface of its body is whitiſh; the ſides a little yellowiſh, variegated with deeper ſpots; the upper part of its head and body variegated with ruſty and dun gray; the ſuperior coverts of its tail yellowiſh, and its quills blackiſh, edged with gray, except the outermoſt, which are white; the wing quills gray, finely edged with a blacker colour: the ſuperior coverts are of the ſame gray, bordered with ruſty; the bill and feet leaden gray.

This bird was ſent from Siberia, where it is ſtill not common.—The navigator John Wood ſpeaks of ſmall birds like larks ſeen at Nova Zembla:

Zembla*: these are probably of the same species, since both prefer an arctic climate.—Lastly, in the *Fauna Russica* I find the *Alauda Tungustica aurita*, or the crested lark of Tunguse, a nation which borders on Siberia. But we still want observations to assign these birds their true place.

Total length, five inches and three quarters; the bill, six or seven lines; the nail, five lines and a half; the tail two inches, composed of twelve quills, and exceeding the wings an inch.

* Hist. Gen. des Voyages, t. xv. p. 167.

FOREIGN

FOREIGN BIRDS

WHICH ARE RELATED TO THE LARK.

I.

THE RUFOUS LARK.

La Variole, *Buff.*
Alauda Rufa, *Gmel.*

COMMERSON brought this beautiful little bird from the country watered by the *de la Plata*. The upper ſurface of its head and body is blackiſh, prettily variegated with different rufous tints: the fore part of its neck is mailed with the ſame; its throat, and all the under ſurface of its body, whitiſh; the quills of its tail brown, the eight middle ones edged with light rufous, and the two outer pairs edged with white; the great quills of the wings gray, and the middle ones brown, all edged with ruſty colour; the bill brown, grooved near the point; the feet yellowiſh.

Total length, five inches and a quarter; the bill, eight lines; the *tarſus*, ſeven or eight lines; the hind toe, three lines, and its nail four lines; the tail, twenty lines, ſomewhat forked, con-

 ſiſting

fifting of twelve quills, exceeding the wings by an inch. [A]

[A] Specific character of the *Alauda Rufa*: "Its tail quills are brown; the eight mid ones rufty at the edge; the outermoft white at the edge."

II.

THE CINEREOUS LARK.

La Cendrille, *Buff.*
Alauda Cinerea, *Gmel.*

I HAVE feen the figure of a lark from the Cape of Good Hope, in which the throat and all the under part of the body were white, the upper part of the head rufous, and a fort of cap with a border of white ftretching from the bafe of the bill beyond the eyes: on each fide of the neck was a rufous fpot edged with black above; the upper furface of the neck and body, cinereous; the fuperior coverts of the wings, and their middle quills, gray; the large quills black, and fo were thofe of the tail.

Total length, five inches; the bill, eight lines; the nail of the hind toe ftraight and pointed, and equal to the toe; the tail, eighteen or or twenty lines, and projecting nine lines beyond the wings.

Is there any analogy between this bird and the cinereous lark which Dr. Shaw saw in plenty near Biserta, which is the ancient Utica? Both these birds are natives of Africa; but the distance is prodigious from the coasts of the Mediterranean to the Cape of Good Hope; and the cinereous lark of Biserta is not sufficiently known to permit us to decide its species. May it not be related to the Senegal lark? [A]

[A] Specific character of the *Alauda Cinerea:* "It is cinereous; its belly and vent white; the quills of its wings and tail brown, the outermost externally white near the tip."

III.

THE AFRICAN LARK.

Le Sirli du Cap de Bonne Esperance, *Buff.*
Alauda Africana, *Gmel.*

IF this bird seem to differ from the larks by the curvature of its bill, it approaches them still more by the length of its hind nail.

All the upper surface is variegated with shades of brown, rufous, and white; the coverts of the wings, their quills, and those of the tail, brown

brown edged with whitiſh, ſome having a double border, the one whitiſh and the other ruſty; all the inferior part of the body whitiſh, ſprinkled with blackiſh ſpots; the bill black, and the feet brown.

Total length, eight inches; the bill, one inch; the *tarſus*, thirteen lines; the hind toe, four lines, and its nail ſeven lines, ſtraight and pointed; the tail, about two lines and a half, conſiſting of twelve quills, and projecting eighteen lines beyond the wings.

THE

No. 119

THE CRESTED LARK.

THE

CRESTED LARK.

Le Cochevis, ou la Grosse Alouette Huppée, *Buff.*
Alauda Cristata, *Linn. Gmel. Brun. Kram. Will. Klein. Briss. & Brown.*
Lodola Capelluta, *Olin. & Zinn.* *

THE crest is, according to Belon, composed of four large feathers; but Olina says there are four or six, and others increase the number even to twelve †. Naturalists agree no better with regard to its position and structure: some assert that it is always erect ‡; others allege that the bird raises or depresses it, con-

* Aristotle calls it Κορυδαλος λοφον εχουσα, i. e. *the Helmet-lark having a crest.* Pliny terms it *Galerita*, and Varro *Galeritus*; both derived from *Galerus*, a *furred cap*, whose primitive is *Galea*, a *helmet.*

In Italian, it has the names *Lodola Capelluta, Capelina, Covarella*, and *Ciperina.*

In German, it is called *Heide-lerche, Baum-lerche, Holtz-lerche*, or *the Heath-lark, the Tree-lark, the Wood-lark, the Way-lark.*

In Danish, it is termed *Top-laerka* and *Vei-laerke.*

The French appellation *Cochevis* is a contraction for *Coq-visage*, or *Cock's-face*; because the tuft resembles somewhat a cock's comb.

† Willughby. ‡ Turner.

tracts or expands it, at pleasure *: nor can we decide whether this difference ought to be imputed to the climate, as Turner hints, or to the age, the sex, or other modifying causes.

The crested lark is a bird of little timidity, says Belon, which seems pleased at the sight of man, and begins to sing when he comes near it. It lives in the meadows and fields, on the sides of ditches and the backs of furrows: it is often seen at the margin of water, and on the high-roads, where it picks its subsistence from the horse-dung, especially in winter. Frisch tells us, that it is found also in the skirts of the woods, perched on a tree; but this seldom occurs, and still more rarely is it discovered in the heart of forests. It sometimes rests on house-tops, the walls of abbeys, &c.

This lark, though not so common as the sky-lark, is however spread through the most of Europe, if not in the northern parts. It is found in Italy, according to Olina; in France, according to Belon; in Germany, according to Willughby; in Poland, according to Rzacynski; and in Scotland, according to Sibbald: but I doubt whether it inhabit Sweden, since Linnæus does not mention it in his *Fauna Suecica*.

The crested lark changes not its abode in winter; but Belon was hasty in suspecting the

* Willughby and Brisson.

text

text of Ariſtotle, ſince it is only ſaid that bird conceals itſelf in that ſeaſon, and ſurely it is leſs ſeen in winter than in ſummer *.

The ſong of the males is loud, yet ſo mellow and pleaſant, that a ſick perſon can bear it in his chamber †.—In order to make them ſing at all times indiſcriminately, they are ſhut up in cages: their warble is ordinarily accompanied by a quivering of the wings. They are the firſt to hail the return of ſpring, and to chant the appearance of the morning, particularly when the air is calm and clear; and ſometimes they warble during the night ‡: for fine weather exhilarates and inſpires congenial notes; but clouds and rain oppreſs their ſpirits, and occaſion languor and gloomy ſilence. They generally ſing till the end of September.—As theſe birds are with difficulty habituated to confinement, and as they live only a ſhort time in the cage, it is proper to ſet them at liberty each year in the end of June, which is the time when they ceaſe to ſing; and to truſt to the catching of others the following ſpring. And though the bird be diſmiſſed, its muſic may be preſerved; for, if it be put beſide a young ſky-lark or a young canary, they will ſoon learn to imitate it §.

* Φωλεῖ γαρ . . . . ϗ κορυδος. *Hiſt. Anim.* lib. viii. 16.
† Traité du Serin, p. 43.
‡ Friſch. § Idem.

But besides the excellence of his warble, the male is distinguished by the strength of his bill, the bigness of his head, and by a larger share of black on his breast *. He conducts his amours in the same way as the sky-lark; only, by reason of the smaller numbers, he is obliged to describe a more spacious circle in the air.

The female constructs her nest like the common lark, but ofteneſt near the highways: she lays four or five eggs, and takes little concern in hatching them. It is even said that a very moderate warmth, aided by the sun's rays, is sufficient for the purpose †; but after the young are excluded, they awaken her tenderness by their repeated cries, and she then provides for them with a maternal affection till they are flown.

Frisch says, that they breed twice a year, and prefer to nestle in junipers: but this remark is chiefly adapted to the climate of Germany.

The early education of the young succeeds easily, but they are always more difficult to be supported afterwards; and it is uncommon, as I

* Olina.

† As these nests are made on the ground, some ignorant credulous person may have seen a toad beside them, or even on the eggs; and hence the fable, that the crested and some other species of larks entrusted the toads with the business of incubation.

have

have asserted on the authority of Frisch, to preserve them alive in the cage for a whole year. And even when we give them the food that agrees best with them, which is ants' eggs, ox and sheep's hearts minced down, bruised hemp-seed and millet, we must be careful, in introducing the little balls into the throat, not to derange the tongue, which might prove fatal.

Autumn is the proper season to lay snares for these birds; and then great numbers in plump condition are caught on the verge of the forests. Frisch observes, that they may be decoyed by the call, which the sky-larks are proof against. The other differences are these: The crested lark never consorts in flocks; its plumage is less varied and more white; its bill is longer, and its tail and wings shorter; it does not mount so high in the air, is less able to struggle with the winds, and returns sooner to the ground. In other properties, the two species are alike, and even with regard to the period of their lives, when not abridged by the constraints of slavery.

Upon the whole, it appears that, of all the larks, the crested, notwithstanding the attachment it is said to shew to man, has the most independent disposition, and recoils from the domestic state. It lives solitary, to escape perhaps the restraints of the social condition.—

 How-

However, it can ſorget its native warble *, and is ſoon taught other airs, which it repeats without blending or confuſion †.

In the ſubject obſerved by Willughby, the tongue was broad and rather forked, the *cæca* extremely ſhort, and the gall-bladder of a dull blueiſh green, which that naturaliſt attributes to ſome accidental cauſe.

Aldrovandus gives the figure of an aged creſted lark. Its bill was white round the baſe; the back cinereous; the under ſide of the body whitiſh, and alſo the breaſt whitiſh, though dotted with brown; the wings almoſt entirely white, and the tail black.

The creſted lark has other enemies beſides man: the ſmalleſt of the rapacious birds attack it, and Albertus ſaw one devoured by a raven. It dreads its ravenous foes ſo much as to throw itſelf at the mercy of the fowler, or remain motionleſs in a furrow till it be caught by the hand.

* The creſted lark is the only one perhaps that may be inſtructed in the ſpace of a month. It repeats an air whiſtled to it, even when ſleeping with its head under its wing; but its voice is very feeble.

† "The creſted lark can learn many airs perfectly, which the canary cannot. . . . Beſides, it retains nothing of its native warble. . . . And this the canary can never lay aſide." *Traité du Serin de Canarie*, p. 43, edit. 1707.

Total

Total length, fix inches and three quarters; the bill, eight or nine lines; the hind toe with the nail the longeft of all, and meafuring nine or ten lines; the alar extent, ten or eleven inches; the tail two inches and a quarter, compofed of twelve quills, and exceeding the wings by thirteen lines. [A]

[A] Specific character of the Crefted Lark, *Aláuda Criftata*: "Its tail quills black, the two outermoft white at their exterior edge; its head crefted; its feet black." Its egg is afh-coloured, with numerous dingy brown fpots.

THE

LESSER CRESTED LARK.

Le Lulu, ou la Petite Alouette Huppée.
Alauda Nemorosa, *Gmel.*
Alauda Cristata Minor, *Ray. Will. & Briss.*
Alauda Cristatella, *Lath. Ind.**

THE chief difference between this and the preceding consists in its inferior size: its plumage is also lighter; its feet reddish; and its cry, which is disagreeable, is never uttered except when it flies †. It is likewise distinguished by its mimicking oddly the songs of other birds ‡: they are not solitary, but rove through the fields in bodies §: the tuft consists of feathers proportionally longer than in the crested lark ||.

This bird is found in Italy, in Austria, in Poland, in Silesia, and in the northern counties of England, as in Yorkshire, &c. but its name appears not in the catalogue of birds that inhabit Sweden.

* Gesner says that its name *Lulu* is formed in imitation of its cry, *lu, lu, lu.*

† Aldrovandus. ‡ Gesner. § Aldrovandus. || I.

It

It frequents heaths, commons, and even woods, whence the German name *Wald-lerche:* in ſuch ſituations it builds its neſt, and hardly ever among corn.

In the rigours of winter, and particularly when the ground is covered with ſnow, it reſorts to the dunghills, and picks up its food about the barns. It alſo haunts the highways, and certainly for the ſame reaſon.

According to Longolius, it is a bird of paſſage, which remains in Germany all winter, and leaves that country about the time of the equinox.

Geſner mentions another creſted lark, of which he ſaw only a drawing, and which had only a ſlight variation of plumage, there being more white round the eyes and the neck, and below the belly. But this might be owing to age, as in the inſtance of the former article; or it might ariſe from ſome accidental cauſe: the difference is undoubtedly inſufficient to conſtitute a diſtinct ſpecies. I muſt however obſerve, that in Geſner's figure the hind nail is not ſo long as uſual in the larks. [A]

[A] Specific character of the Leſſer Creſted Lark, *Alauda Nemoroſa*: "Its tail quills are black, the two outermoſt white on their exterior edge; its head creſted; its feet red." Its egg is light red, with a few ferruginous ſpots.

THE

THE

UNDATED LARK.

La Coquillade, *Buff.*
Alauda Undata, *Gmel.*

THIS is a new ſpecies, which M. Guys ſent to us from Provence. It appears to be related to the creſted lark; for it has a ſmall ſupine tuft, which it probably can elevate at pleaſure. It is properly the bird of the morning, ſince it begins with the earlieſt dawn, and ſeems to rouſe the other birds. According to Guys, the male does not leave his mate when ſhe hatches; and when the one is employed in ſeeking their food, which conſiſts of caterpillars, graſhoppers, and even ſnails, the other keeps watch, to give the ſignal when danger threatens.

The throat and all the under ſide of the body are whitiſh, with ſmall blackiſh ſpots on the neck and breaſt; the tuft feathers blackiſh, edged with white; the upper ſide of the head and body, variegated with blackiſh and light rufous: the great coverts of the wings tipt with white; the quills of the tail and of the wings, brown edged with light rufous, except ſome in the

the wings that are edged or tipt with white: the bill is brown above, and whitiſh below; the feet yellowiſh.

Total length, ſix inches and three quarters; the bill, eleven lines, and pretty ſtrong; the *tarſus*, ten lines; the hind toe, nine or ten lines, the nail (which is eight lines) included; the tail two inches, exceeding the wings ſeven or eight lines.

Sonnerat brought a lark very like this from the Cape of Good Hope; the only difference being, that it had no creſt, that the colour of the under ſide of its body was more yellowiſh, and that none of the quills of the tail or wings was edged with white.—It was perhaps a female, or a young bird of the year's hatch.

In Haſſelquiſt's Travels, a Spaniſh lark is mentioned, which that naturaliſt ſaw in the Mediterranean the inſtant it left the ſhore; but he takes no farther notice of it, and I can find no ſpecies deſcribed by authors under that appellation. [A]

[A] Specific character of the *Alauda Undata*: "Its tail quills are brown, tawny at the edge; its feet yellowiſh; the feathers of its creſt black, edged with white."

FOREIGN BIRD

RELATED TO THE CRESTED LARK.

THE SENEGAL LARK.

La Grisette, ou le Cochevis du Senegal.
Alauda Senegalensis, *Gmel.*

WE owe to Brisson almost all we know of this foreign bird. It has a sort of tuft, consisting of feathers that are longer than those which cover the rest of the head: it is nearly as large as the sky-lark. It inhabits Africa, and perches on the trees which grow on the banks of the Niger: it is also seen in the island of Senegal. The upper side of its body is variegated with gray and brown; the superior coverts of the tail, rusty gray; the under surface of its body whitish, with small brown spots on the neck; the quills of its wings brown gray, edged with gray; the two middle ones of its tail gray; the lateral ones brown, except the outermost, which is rusty white, and the next, which is edged with the same: the bill is horn colour; the feet and nails gray.

I have seen a female, whose tuft was laid back

back as that of the male, and variegated, as well as the head and the upper ſide of the body, with brown ſtreaks on a ruſty ground: the reſt of the plumage conformed to the preceding deſcription. The bill was longer, and the tail ſhorter.

Total length, ſix inches and a half; the bill, nine lines and a half; the alar extent, eleven inches; the hind toe, including the nail, equal to the mid toe; the tail, two inches two lines, rather forked, and conſiſting of twelve quills; it exceeds the wings by ſix or ſeven lines. [A]

[A] Specific character of the *Alauda Senegalenſis*: "Its two middle quills are gray, the reſt brown; the outermoſt rufous brown on the exterior ſide; the head ſomewhat creſted."

THE

NIGHTINGALE.

Le Rossignol, *Buff.*
Motacilla-Luscinia, *Linn. & Gmel.*
Luscinia, *Will. Ray. Sibb. Briss* &c.
Sylvia-Luscinia, *Lath. Ind.**

TO every person whose ear is not totally insensible to melody, the name of nightingale must recall the charms of those soft evenings in spring, when the air is still and serene, and all nature seems to listen to the songster of the grove. Other birds, the larks, the canaries, the chaffinches, the petty-chaps, the linnets, the goldfinches, the blackbirds, the American mocking-birds, excel in the several parts which they perform †: but the nightingale combines

* In Greek, the nightingale has been styled by pre-eminence Αηδων, from αειδω, to sing: in Hebrew, its name is supposed to be *Trachmas*: in Arabic, *Enondon*, *Audon*, *Odorbrion*. Its Latin appellation *Luscinia* is of uncertain derivation; that of *Philomela* is rather poetic, and signifies *a lover of song*: in Italian, *Rossignuolo*, *Uscigniulo*: in Spanish, *Ruissenol*: in German, *Nachtigal*, *Doerling*, *Tagschläeger*: in Dutch, *Nachtegael*: in Danish, *Nattergal*. The English name is formed from the Saxon word *galan*, to sing.

† "I have happened," says Mr. Barrington, "to hear the American

N.o 120

FIG.1. THE NIGHTINGALE. FIG.2. THE REDSTART.

bines the whole, and joins ſweetneſs of tone with variety and extent of execution. His notes aſſume each diverſity of character, and receive every change of modulation; not a part is repeated without variation; and the attention is kept perpetually awake, and charmed by the endleſs flexibility of ſtrains. The leader of the vernal chorus begins the prelude with a low and timid voice, and he prepares for the hymn to nature by eſſaying his powers and attuning his organs †: by degrees the ſound opens and ſwells; it burſts with loud and vivid flaſhes;

American mocking-bird in great perfection. . . . . During the ſpace of a minute, he imitated the wood-lark, chaffinch, blackbird, thruſh, and ſparrow. I was told alſo that he would bark like a dog; ſo that the bird ſeems to have no choice in his imitations, though his pipe comes neareſt to our nightingale, of any other bird I have met with. . . . . Kalm indeed informs us that the natural ſong is admirable; but this traveller ſeems not to have been long enough in America to have diſtinguiſhed what were the genuine notes: with us, mimics do not often ſucceed but in imitations. I have little doubt, however, but that this bird would be fully equal to the ſong of the nightingale in its whole compaſs; but then, from the attention which the *mocker* pays to any other ſort of diſagreeable noiſes, theſe capital notes would always be debaſed by a bad mixture."

Philoſoph. Tranſ. vol. lxiii. p. 2.

† "I have commonly obſerved," ſays Mr. Barrington, "that my nightingale began ſoftly, like the ancient orators; reſerving its breath to ſwell certain notes, which by this means had a moſt aſtoniſhing effect, and which eludes all verbal deſcription."

it

it flows with ſmooth volubility; it faints and murmurs; it ſhakes with rapid and violent articulations: the ſoft breathings of love and joy are poured from his inmoſt ſoul, and every heart beats uniſon, and melts with delicious languor. But this continual richneſs might ſatiate the ear. The ſtrains are at times relieved by pauſes *, which beſtow dignity and elevation. The mild ſilence of evening heightens the general effect, and not a rival interrupts the ſolemn ſcene.

The nightingale excels all birds in the ſoftneſs and mellowneſs, and alſo in the duration of its warble, which ſometimes laſts without interruption twenty minutes. Barrington reckons ſixteen diſtinct notes between the higheſt and loweſt keys, and finds that its ſong fills a circle of a mile in diameter, which is equal to the power of the human voice. Mr. J. Hunter diſcovered that the muſcles of the larynx are ſtronger in this ſpecies than in any other, and even that they were ſtrongeſt in the male, which alone ſings.

Ariſtotle, and after him Pliny, affirms that

* The Engliſh bird-catchers, we are told by Mr. Barrington, give theſe names to ſome of the nightingale's notes: *Sweet*, *Sweet jug*, *Jug ſweet*, *Water bubble*, *Pipe rattle*, *Bell pipe*, *Scroty*, *Skeg*, *Skeg*, *ſkeg*, *Swat ſwat ſwaty*, *Whitlow*, *whitlow*, *whitlow*, from ſome diſtant affinity to the words.

the

the ſong of the nightingale laſts in its full vigour for fifteen days and nights, without intermiſſion, at the time when the trees expand their foliage. But this muſt be reſtricted to the wild nightingales, and even theſe are not always ſilent before and after the period aſſigned; indeed, they ſing then with moſt animation and conſtancy. They generally begin in the month of April, and ceaſe not until June, about the time of the ſolſtice. Their ſilence is greateſt when the young are excluded from the eggs, and abſorb their attention in providing food. The tame nightingales ſing during ſix months, and their warble has not only more extent, but is more perfectly formed: whence Barrington infers, that in this ſpecies, as well as in many others, the male does not chant with the view of amuſing the female, or of relieving the tedium of incubation; a concluſion which is entirely juſt and well founded. The female hatches from inſtinct; ſhe is prompted by a paſſion ſtronger than love itſelf; and, in that occupation, ſhe muſt feel a pleaſure, which, though we cannot conceive it, muſt be exquiſite, and needs no ſolace. The male is the moſt powerfully ſtimulated to court the joys of love, and warble his amorous tales; and the opening fluſh of ſpring inflames his ardent deſires. When the nightingale is confined, his wants are ſupplied and anticipated, and he enjoys the vernal mildneſs through the greater part of the year; every thing concurs to foſter his paſſion, and the melody of his ſong ceaſes only for a ſhort interval. Such is alſo the caſe with thoſe which are caught after they are grown up; within a few hours they often reſume their warble, in all its richneſs of expreſſion; no longer is it limited by the tranſient ſeaſons. But the love of liberty is not extinguiſhed in their boſom: for the firſt week, they can hardly ſupport their forlorn condition; they muſt be pinioned and fed. However, their paſſion for warbling in the end ſurmounts every deſponding recollection. The ſong of other birds, the muſic of inſtruments, the accents of a ſweet and ſonorous voice, powerfully awaken their languid powers. They are eagerly attracted by ſweet ſounds, and ſeem particularly fond of *duos*; which ſhews that they are not inſenſible to the effects of harmony. But they are not ſilent auditors; they ſtrike the uniſon, and ſtrive to eclipſe their rivals. It is ſaid that a nightingale dropped dead at the feet of a perſon who ſung; another fretted, ſwelled its throat, and murmured diſcontent whenever a canary, which was beſide it, began to chant, till at laſt, by its menaces, it ſilenced its competitor*.—Real ſuperiority is not always exempt

* *Note of M. de Varicourt, Advocate.* M. le Moine, treaſurer of France, at Dijon, who takes pleaſure in raiſing nightin-

exempt from jealousy. May not this desire of excelling be the reason why they choose advantageous situations, and delight to sing to the returning echo?

The nightingales differ much in the quality of their song; in some it is so inferior, that they are reckoned not worth keeping. It is even said, their warble is not the same in every country: the bird fanciers in England prefer the nightingales of Surry to those of Middlesex; as they value most the chaffinches from Essex, and the goldfinches from Kent. This diversity of song has been properly compared to the different dialects of the same language. It may arise from accidental causes: a nightingale happens to hear the sweet music of some other birds, and, in the glow of emulation, improves his own; he communicates the melody to his young *; and thus it is transmitted, with various alterations, through the succeeding races.

nightingales, remarked also that his pursued bitterly a tame canary, that was kept in the same room, when it approached their cage. But this jealousy turns sometimes into emulation; for nightingales have been observed to sing better than others, merely because they heard birds whose strains were inferior to theirs. *Certant inter se, palamque animosa contentio est: victa morte finit sæpe vitam.* Plin. lib. x. 29. They have been imagined to sing duos, with the interval of a third.

* Pliny relates, that the parent is careful in instructing his young. Lib. x. 29. & lib. iv. 9.

 After

After the month of June, the nightingale's warble is gone; a raucous croaking remains, and the ſweet Philomela is no longer recognized. No wonder that, in ancient Italy, it then received another name *. In regard to ſong, it is quite a different bird, and even the colours of its plumage are ſomewhat altered.

In the nightingales, as in other ſpecies, there are females which enjoy ſome prerogatives of the male, and particularly participate of his ſong. I ſaw a female of that ſort which was tame; her warble reſembled that of the male, yet neither ſo full nor ſo varied: ſhe retained it until ſpring, when, reſuming the character of the ſex, ſhe exchanged it for the occupation of building her neſt and laying her eggs, though ſhe had no mate. It would ſeem that in warm countries, as in Greece, ſuch females are pretty common, both in this ſpecies, and many others at leaſt this is implied in a paſſage of Ariſtotle †.

A muſician, ſays Friſch, ought to ſtudy the ſong of the nightingale, and write it. This is what the jeſuit Kircher formerly attempted, and which has been lately tried by Barrington.

* Ariſtotle, *Hiſt. Anim.* lib. ix. 49.

† "Some males ſing like their females, as appears in the genus of the nightingales; but the female gives over ſong when ſhe hatches." *Hiſt. Anim.* lib. iv. 9.

The enthuſiaſts of muſic imagine, that the ſong of the nightingale contributes more than warmth to quicken the chick in the egg.

Yet

Yet the latter owns that he could not ſucceed; for though the notes were played by an excellent performer on the flute, they bore no reſemblance to the nightingale's warble. Mr. Barrington ſuſpects the difficulty to ariſe from the impoſſibility of marking the muſical intervals. Their meaſures are ſo varied, their tranſitions ſo inſenſibly blended, the ſucceſſion of their tones ſo wild and irregular as to ſoar beyond every conſtraint of method: but ſtill more difficult it would be, to imitate with a dead inſtrument the expreſſion of the nightingale, his tender ſighs, and his melting accents. The animated organ of voice can alone aſpire to the muſic of nature; and the rare accompliſhment of a ſoft, clear, flexible, ſonorous tone, of a delicate and diſcerning ear, and of an exquiſite ſenſibility, will ſometimes ſucceed. I knew two perſons, who, though they had not noted a ſingle bar, imitated the whole extent of the nightingale's warble, and ſo accurately as to deceive the hearer: they were two men, who rather whiſtled than ſung: the one whiſtled ſo ſoftly, that it was impoſſible to perceive the motion of his lips; the other blew with more force, and his attitude was ſomewhat conſtrained, though his imitation was ſtill perfect. A few years ago, there was a man at London who attracted the nightingales with his ſong; inſomuch that they alighted on

 him,

him, and allowed themſelves to be caught by the hand*.

But though few can imitate the nightingale, every perſon is eager to enjoy its ſong. It is a domeſtic of a capricious humour, which neither cheerfulneſs nor affection can direct. It muſt be treated with tenderneſs, and ſupplied abundantly with the food it likes: the walls ſhould be painted with verdure, and ſhaded and enriched with foliage; moſs ſhould be ſtrewed beneath its feet; it muſt be guarded againſt cold, and from troubleſome viſits †; and every illuſion muſt be employed to diſpel the gloom of captivity. Such precautions will ſucceed. If an old one is caught in the beginning of ſpring, it will ſing in ſeven or eight days, or even ſooner ‡, and will renew its warble every year, in the month of May and the end of December. If young ones of the firſt hatch are educated, they will begin to chant as ſoon as they can eat by themſelves; their voice will riſe and form by degrees, and attain its full perfection about the

* *Annual Regiſter*, for 1764. *Aldrovandus*, p. 783. "There are perſons, who, putting water into tranſverſe reeds, and blowing into an aperture, imitate the ſound perfectly." *Pliny*, lib. x. 29.

† It is even recommended to clean it ſeldom when it ſings.

‡ Thoſe which are taken after the 15th of May ſing ſeldom the reſt of the ſeaſon. Thoſe which ſing not in a fortnight will never ſing, and are often females.

end

end of December. Their warble is much ſuperior to that of the wild nightingales, and will flow each day of the year, except in the ſeaſon of moulting. They will appropriate the notes of other birds, through imitation or rivalſhip*, and they will even chant the ſtiff airs of a nightingale-pipe. They may be inſtructed even to ſing by turns with a chorus, and repeat their couplet at the proper time. They may be alſo taught to ſpeak any language. The ſons of the emperor Claudius had ſome nightingales that ſpoke Greek and Latin †. But what Pliny ſubjoins is more marvellous; that theſe birds prepared every day new expreſſions, and even of conſiderable length, with which they entertained their maſters ‡. The arts of flattery might work upon the underſtandings of young princes; but a philoſopher like Pliny ought not to have credited ſuch a ſtory, nor have publiſhed it under the ſanction of his name. Several authors accordingly, reſting on the authority of the Roman naturaliſt, have amplified the marvellous tale. Geſner, among others, quotes a letter from a perſon of credit (as we ſhall find), who mentions two nightingales belonging to an innkeeper at Ratiſbon which paſſed the night diſ-

* Pliny, lib. iv. 9. & lib. x. 29.

† Pliny, lib. x. 42.

‡ Theſe young princes were Druſus and Britannicus.

courſing in German on the political intereſts of Europe; on the events already happened, and on thoſe which might be expected, and which actually took place. It is true that the author of the letter endeavours to render the ſtory more probable, by telling us that the birds only repeated what they heard of ſome officers, or deputies from the Diet, who frequented the tavern; but ſtill the whole is ſo abſurd as to merit no ſerious refutation.

I have ſaid that the old priſoners had two ſeaſons for ſinging; the months of May and December. But here art interferes, and again diverts the courſe of nature. The chamber is darkened, and the birds are ſilent. If the light be reſtored by degrees, it will produce the effect of ſpring, and the nightingales will reſume their warble. If we have a ſufficient number of old ſingers, we may, by ſucceſſive manœuvres, and by haſtening or retarding the time of moulting, be entertained with continual melody. Some of the young birds which are raiſed, ſing during the night; but moſt of them begin to chant at eight or nine of the morning, in the winter ſolſtice, and gradually earlier as the days lengthen.

One would hardly believe that a ſong ſo varied as that of the nightingale is confined within the narrow limits of a ſingle octave; yet this is aſcertained by an obſerver of infor-

mation

mation and taſte*. He remarks indeed, that ſome tones ſhoot like a gleam to the ſecond octave; but theſe are accidental, and occaſioned by violent efforts of voice, as when a perſon happens to blow into a flute with exceſſive force, he produces ſounds uncommonly ſhrill.

Though ſlow in acquiring it, the nightingale is ſuſceptible of an unſhaken attachment. It diſtinguiſhes the ſtep of its maſter, and welcomes his approach with the muſic of joy; and even in the moulting ſeaſon, it idly ſtrains its enfeebled organs to expreſs the warmth of its heart. Sometimes it dies of melancholy at the loſs of its benefactor; and, if it ſurvives the ſhock, it is tardy in forming new attachments †.—Such is even the caſe with ſhy and timid characters; their intimacies are ſlow and difficult, but, once eſtabliſhed, they are ardent and durable.

The nightingales are alſo very ſolitary; they migrate alone, appearing in April and May, and retiring in September ‡. The union between the ſexes in the ſpring ſeems to increaſe their averſion to ſociety: they ſelect certain

* Dr. Remond, who has tranſlated ſeveral pieces of the *Collection Academique*.

† M. le Moine.

‡ In Italy, they arrive in March and April, and retire about the beginning of November. In England, they arrive in April and May, and retire in the month of Auguſt.

tracts,

tracts, and oppose the encroachments of others on their territories. But this conduct is not occasioned by rivalship, as some have supposed; it is suggested by the solicitude for the maintenance of their young, and regulated by the extent of ground necessary to afford sufficient food. The distances between their nests are much smaller in the rich counties, than in others which reluctantly yield a penurious supply.

They begin to build their nest about the end of April, or the opening of May. The materials are leaves, rushes, stalks of coarse grass, and the inside is lined with small fibres and roots, horse-hair, and a sort of bur. They are placed in a favourable aspect, turned somewhat to the east, near water, and commonly on the lowest branches of shrubs, as gooseberries, white thorns, sloes, elm-hedges, &c. or in a tuft of grass, and even on the ground under bushes. And hence their eggs and their young, and sometimes the mother, are often devoured by dogs, foxes, pole-cats, weasels, adders, &c.

In our climate, the female usually lays five eggs*, of an uniform greenish brown, except that the brown predominates at the obtuse end, and the green at the acute. She hatches alone, and never leaves her station but for food, and

* Aristotle says five or six; which may be true in the farm climate of Greece.

then

then only in the evening, and when hard preſſed with hunger: during her abſence, the male ſeems to caſt an eye on the neſt. In eighteen or twenty days the young begin to burſt their ſhells, and the number of the cock birds is generally double that of the hens. And hence, when in April a cock is caught, the hen ſoon finds another mate, the loſs of which is ſupplied by a third, and ſo on; inſomuch that the ſucceſſive ſeizure of three or four males has little effect on the multiplication of the brood. The hen diſgorges the food for the young, as in the canaries; and the cock aſſiſts. He now ceaſes to warble, and is totally abſorbed in the concerns of his family: and even during incubation, it is ſaid, he ſeldom ſings near the neſt, leſt he diſcover it; but if a perſon approach it, his paternal ſolicitude drowns the ſuggeſtions of prudence, and his ſhrieks only increaſe the danger.—In leſs than a fortnight the young are fledged, and at this time we ought to remove thoſe intended to be trained. After they are flown, the parents make a ſecond hatch, and then a third; but the laſt fails, if the cold ſets in early. In hot countries, they breed even four times annually; but the late hatches are always more ſcanty.

Man, who is never ſatisfied till he can uſe and abuſe what he poſſeſſes, has contrived to make the nightingales breed in their priſon. The

The great obſtacle to his plans was their ardent love of liberty; but he has diverted this original bent by foſtering more violent paſſions, the impulſe for propagation, and the attachment to offspring. A pair of nightingales are let looſe into a large volary, or rather in a corner of a garden planted with ivy, hornbeams, and other ſhrubs, and incloſed with nets. This is the eaſieſt and ſureſt method; but there is another which alſo may be employed. The cock and hen are put into ſeparate cages in a darkiſh cloſet, and are fed every day at the ſame hours; the cages are ſometimes left open, that they may become acquainted with their apartment, and in April they are entirely opened, and not ſhut again; at which time the birds are ſupplied with the materials uſual for conſtructing their neſts, ſuch as oak-leaves, moſs, plucked dog-graſs, deer's wool, horſe hairs, earth and water; but the water muſt be removed when the female hatches *. A contrivance has alſo been uſed to ſettle nightingales in places which they never viſited: the brood are caught with the parent birds, and conveyed to a ſpot which reſembles moſt their former haunt: the cock and hen are placed in two cages near the neſt of young, and

* Traité du Roſſignol, p. 96.

the

the cages are artfully opened while the person remains concealed: the parents run to the cries of their nestlings, and foster and continue to educate them: and it is said they will return to the same part the succeeding year *; but it must afford them conveniences and proper food †.

If a person would raise young nightingales, he should prefer those of the first hatch, and give them what instructors he thinks proper; but the best, in my opinion, are other nightingales, especially the best singers.

In the month of August, both the old and young nightingales emerge from the woods, and haunt the bushes, the quick-set hedges, and the new ploughed fields, where they find greater plenty of worms and insects: perhaps this general movement is only preparatory to their approaching flight. They winter not in France, nor in England, Germany, Italy, Greece, &c. ‡; and as we are assured that no nightingales occur in Africa §, they would seem to retire into

* Traité du Rossignol, p. 96.

† "When a place affords abundance of food and convenience for nestling, one had better catch or destroy the nightingales, since others will always return." *Frisch.*

‡ The nightingale disappears in autumn, and appears not again till spring, says Aristotle. *Hist. Anim.* lib. v. 9.

§ *Traité du Rossignol,* p. 21. Indeed the voyager Le Marie speaks of a nightingale at Senegal, but which sings not so well as ours.

Asia.

Asia *. And this opinion is the more probable, as they inhabit Persia, China, and even Japan, where they are highly esteemed, since the fine warblers are sold, it is said, for twenty cobangs †. They are spread generally through Europe, even to Sweden and Siberia ‡, where they chant delightfully; but there are countries in Europe, as well as in Asia, which are not suited to them, and where they never fix their abode: for instance, in Bugey as far as the heights of Nantua; a part of Holland, Scotland, Ireland §, the northern part of Wales, and even of all England except Yorkshire ‖; the territory of the Daulians

* Olina. They are found among the willows and olives of Judea. *Hasselquist*.

† Kæmpfer. The cobang is equal to forty taels, and the tael worth about half a crown; so that the twenty cobangs amount to an hundred pounds sterling. Nightingales were much dearer in Rome, as we shall see in the article of the white nightingale.

‡ Gmelin speaks with raptures of the charming banks of a rivulet in Siberia, called *Beressouka*, and of the warble of the birds heard there, among which the nightingale holds the first rank.

§ Aldrovandus. I know that the fact has been doubted with regard to Ireland, Scotland, and Holland; but these assertions must not be strictly interpreted. No more is meant than that the nightingales are extremely rare in those countries; and the case must be so where there are few woods or bushes, little heat, few insects, few fine nights, &c.

‖ Mr. Pennant's words are these: "It is not found in North Wales; or in any of the English counties north of it, except

Daulians near Delphi; the kingdom of Siam, &c. * They are univerſally known to be birds of paſſage, and thoſe which are confined appear impatient and reſtleſs in the ſpring and autumn, eſpecially during the night, their uſual periods of migration: a proof that their peregrinations are occaſioned by incitements very different from the want of food, or the deſire of warmth.

This bird is peculiar to the old continent; and though the miſſionaries and voyagers ſpeak of the nightingales of Canada, that of Louiſiana, that of the Antilles, &c. the latter is only a kind of mocking-bird; and that of Louiſiana is the ſame, ſince, according to Dupratz, it is found in Martinico and Guadaloupe; and it is manifeſt, from Father Charlevoix's account, that the one of Canada is either not a nightingale at all, or is of a very degenerate ſort †. It is indeed poſſible that the nightingale, which

except Yorkſhire, where they are met with in great plenty about Doncaſter. It is alſo remarkable, Mr. Pennant adds, that this bird does not migrate ſo far weſt as Cornwall; a county where the ſeaſons are ſo mild, that myrtles flouriſh in open air during the whole year." T.

* Voyages de Struys, t. i. p. 53.

† "The Canada nightingale," ſays this miſſionary, "is nearly the ſame with reſpect to figure, but has not half the ſong." *Nouvelle France*, t. iii. p. 157.

visits

visits the northern parts of Europe and Asia, may have traversed the narrow seas, which, at those high latitudes, divide the two continents; or it might have been swept across by a violent wind, or carried by some ship: and as the cold, raw, unfertile state* of the country has been found to be unfavourable to the song of birds, it would lose the sweetness of its melody, as the nightingale's warble in Scotland † is inferior to what is heard in the groves of Italy. This seems to have been really the case; and our nightingale has been transplanted into Canada; for the imperfect account of Charlevoix has since been confirmed by a physician ‡ residing at Quebec, and by some travellers.

As these birds, at least the males, pass the nights of spring in singing, the ancients believed that during that season they never slept §;

* I know that there are many insects in America; but most of them are so large and so well armed, that the nightingale, far from preying upon them, would scarcely be able to defend himself against their attacks.

† Aldrovandus, who cites Petrus Apponensis. This bird must sometimes, therefore, appear in Scotland.

‡ This physician wrote to M. Salerne, that our nightingale is found in Canada, as well as here, in the season. It occurs also in Gaspesia, according to Father Le Clerc, but does not sing so well.

§ Hesiod and Ælian.

and

and hence they inferred that the fleſh had a ſoporific quality, and imagined that its heart and eyes laid under a perſon's head would keep him awake. This notion ſpread; and being transferred into the arts, the nightingale became the emblem of vigilance. But the moderns, who have obſerved with greater accuracy, find that in the ſeaſon of love the nightingales ſleep during the day, and that in winter their diurnal ſlumbers precede the reſuming of their warble. They even dream, and their viſions have the complexion of their character *; for they hum their airs with a low voice.—Many fables have been propagated with regard to the nightingale, as uſual with whatever is celebrated. In the midſt of its warble, it is attracted by the fixed looks of a viper, or, according to others, of a toad, and is gradually deprived of its voice; and at laſt, yielding to the faſcination, it ſinks into the open mouth of the reptile. It has been ſaid that the parents ſelect from their young thoſe which diſcover talents, and either deſtroy the reſt, or allow them to periſh of want. (We muſt ſuppoſe that they ſave the females.) It has been alleged that they chant better when perſons liſten to them than when alone. All theſe errors originate from our proneneſs to aſcribe to animals our own weakneſſes, paſſions, and vices.

* Traité du Roſſignol.

Nightingales that are kept in the cage commonly bathe after ſinging. Hebert remarks, that this is the firſt thing they do in the evening, when the candles are lighted. He alſo tells a circumſtance which it may be proper to guard againſt, that they are apt to dart towards the flame; and that an excellent cock bird, happening to eſcape from the cage, flew into the fire, and was burnt to death.

Theſe birds have a ſort of vibrating motion, and by turns raiſe and depreſs their body. All the males which I have ſeen had this ſingular property; but I could not perceive it in a female which I kept two years. In all of them, the tail oſcillates diſtinctly upwards and downwards; which has manifeſtly induced Linnæus to range them with the *motacillæ*, or wagtails.

Nightingales hide themſelves in the thickeſt buſhes: they live upon aquatic and other inſects, ſmall worms, the eggs, or rather nymphs, of ants; they alſo eat figs, berries, &c. But as it would be difficult always to provide ſuch food, ſeveral kinds of paſte have been contrived, which agree very well with them. I ſhall, in the notes, give the receipt of a bird-fancier of my acquaintance *, becauſe it

* M. Le Moine, whom I have already quoted on ſeveral occaſions, gives different paſtes, according to the age of the bird.

it has been found to anſwer, and I have ſeen a nightingale which with this ſubſiſtence reached the age of ſeventeen years. It began to turn gray the ſeventh year; at fifteen, its wing and tail quills were entirely white; its legs, or rather *tarſus*, had much ſwelled, by the exceſſive growth of the ſcales which cover theſe parts; and it had a ſort of knots at the toes, like perſons who are gouty; and it was neceſſary, from time to time, to whet the point of its upper mandible *. But ſtill it had none of the ills of old age; it was ever joyous, and warbled as in its earlier years, and careſſed the hand that fed it.

bird. The firſt is compoſed of ſheep's heart, crumbs of bread, hemp-ſeed, and parſley well pounded, and mixed: it muſt be new-made every day. The ſecond conſiſts of equal parts haſhed omelet and bread-crumbs, with a little chopped parſley. The third is more compound, and requires more care. Take two pounds of lean beef, half a pound of chicken-peaſe, as much of yellow or peeled millet, of the ſeed of white poppy, and of ſweet almonds, a pound of white honey, two ounces of flour, twelve yolks of freſh eggs, two or three ounces of freſh butter, and a gros and a half of ſaffron in powder; dry the whole well; heat it a long time, ſtirring it conſtantly, and reduce it to a very fine duſt, and paſs it through a ſilk ſieve. This powder keeps very well, and ſerves about a year.

* The nails alſo of nightingales that are kept in the cage grow much at firſt, ſo as to become troubleſome from their exceſſive length. I have ſeen ſome which formed a circle of five lines diameter. But in extreme age they diſappear almoſt entirely.

it. We may remark, that this nightingale had never paired. Love ſeems to abridge the period of life; but it fills up the void of time, and accompliſhes the intention of nature. Without it the delightful emotions of the parent would never be known: it prolongs the exiſtence into futurity, and confers immortality on the ſpecies. So rich the compenſation it makes for the loſs of a few days of injoyous and decrepit old age!

It has been found, that heating medicines and perfumes incited the nightingales to warble; that flour mites and dung maggots were the moſt proper to give when they were too fat, and figs when too lean; and that ſpiders ſerved them as a purge. It has been recommended to make them take this purge once a year; a dozen ſpiders is the proper doſe. They ſhould alſo have nothing that is ſalt.

When they ſwallow any indigeſtible ſubſtance, they regorge it in little pellets like the birds of prey; and indeed, though they are ſmall, they merit that appellation. Belon admires *their foreſight in not ſwallowing a ſmall worm before they have killed it*; but it is probable that they only want to avoid the painful feeling which living creatures might occaſion in their ſtomach.

All ſorts of ſnares ſucceed with the nightingale; they are not ſhy, though rather timorous. If

If they be let looſe in a place where there are other birds in cages, they fly directly towards theſe; and this is one method among many others of catching them. The ſong of their companions, the ſound of muſical inſtruments, or of a fine voice, as we have already ſeen, and even cries that are diſagreeable as the mewing of a cat faſtened to the root of a tree and teaſed on purpoſe, ſucceed equally well. They have a ſtupid ſtaring curioſity, and are the dupes of every thing*. They are caught by the call, by lime-twigs in tit-mice traps, and nooſes ſet where the ground is newly ploughed †, where are previouſly ſcattered the nymphs of ants, mites, or whatever reſembles theſe, as ſmall bits of the whites of hard eggs, &c. The gins ought to be made of ſilk ſtrings, and not of packthread, which would entangle their feathers, and occaſion a loſs of ſome, that would retard their ſong. On the contrary, to haſten their moulting, a few feathers ſhould be plucked from the tail, that the new ones may ſprout

* *Avis miratrix*, ſays Linnæus.

† Sometimes they are very numerous in a ſingle diſtrict. Belon witneſſed that, in a village of the foreſt of Ardenne, the ſhepherd boys caught every day each a ſcore, with many other ſmall birds. It was a year of drought, and all the "pools," ſays Belon, "were elſewhere dried up . . . for they live then in the foreſts, where there is ſome moiſture."

the ſooner; for while nature is employed in the reproduction of the plumage, ſhe ſuſpends the ſong.

Theſe birds are delicious food when fat, and in this reſpect they rival the ortolans. In Gaſcony, they are fatted for the table. This reminds us of the whim of the Emperor Heliogabalus, who ate the tongues of nightingales, of peacocks, &c. and the famous diſh of the comedian Æſop, which conſiſted of an hundred birds, all noted for their ſong or their prattle *.

As it is a matter of ſome conſequence not to loſe time in breeding females, many marks have been given to diſtinguiſh the cocks. Their eye, it is ſaid, is larger, their head rounder, their bill longer, and broader at the baſe, eſpecially if viewed from below; the higher plumage more coloured, the belly not ſo white, the tail more feathered, and broader when diſplayed: they begin ſooner to chirp, and their chirp is better ſupported: their anus is more prominent in the love ſeaſon, and they remain long in one place, reſting on one foot, while the hen runs backwards and forwards in the cage. Others add, that the male has two or three quills

* Pliny, lib. x. 51. This diſh was valued at ſix hundred *ſeſterces* (near *five thouſand pounds!*). Aldrovandus alſo ate nightingales, and found them good.

in

in each wing whoſe outer ſurface is black, and their legs, if viewed in a ſide light, appear reddiſh, while thoſe of the female look whitiſh: however, the hen has the ſame motion of the tail; and, when cheerful, ſhe does not walk, but hops like the cock. To theſe we may join the more deciſive differences of internal ſtructure. The males which I diſſected in the ſpring had two very large teſticles of an egg ſhape; the largeſt of the two (for they were not equal) was three lines and a half in length, and two in breadth: the *ovaria* of the females which I obſerved at the ſame time contained eggs of different ſizes, from a quarter of a line to one line in diameter.

The plumage of this bird is much inferior to its warble. All the upper ſide of the body is brown, more or leſs ruſty; the throat, breaſt and belly are of a white gray; the fore part of the neck, of a deeper gray; the inferior coverts of the tail and wings, ruſty white and of a ferruginous caſt in the males; the quills of the wings, of a browniſh gray verging on rufous, and the tail of a brown tinged more with rufous: the bill is brown, and alſo the feet, but with a ſhade of fleſh colour: the ground colour of the feathers is deep cinereous.

It has been alleged, that the nightingales bred in the ſouthern climates have a darker plumage, and thoſe raiſed in the northern countries a whiter. The young cocks are alſo ſaid to be whiter than young hens; and in general the colour of the young ones is the moſt variable before moulting, that is, before the end of July; and it is then ſo like that of the redſtart, that it would be difficult to diſtinguiſh them unleſs by the difference of their chirping *: accordingly theſe two ſpecies are related †.

Total length, ſix inches and a quarter; the bill, eight lines, yellow within, and having a large opening; the edges of the upper mandible ſcalloped near the point; the *tarſus*, an inch; the outer toe connected to the middle one at its baſe; the nails ſlender, the hind one the ſtrongeſt; the alar extent, nine inches; the tail thirty lines, conſiſting of twelve quills, and exceeding the wings ſixteen lines.

Inteſtinal tube, from the ſtomach to the anus, ſeven inches and a quarter; the *æſophagus* near two inches, and dilating into a ſort of glandulous ſac before its inſertion into the gizzard, which is muſcular, and occupies the left ſide of the lower belly, and not covered by the inteſtines, but only by a lobe of the liver: there

* The young cock nightingale calls *ziſcra, ziſcra*, according to Olina; *croi, croi*, according to others. Each perſon has his manner of hearing and expreſſing theſe indeterminate ſounds, which are themſelves ſo variable.

† It is even ſaid that they breed together.

are

are two exceeding ſmall *cæca*, and a gall bladder: the tip of the tongue is furniſhed with threads, and truncated, which was known to the ancients *; and ſeems to have given riſe to the fable of Philomela, whoſe tongue was cut out. [A]

[A] Specific character of the Nightingale, *Motacilla-Luſcinia*: "It is rufous cinereous, its braces cinereous."

VARIETIES of the NIGHTINGALE.

I. THE GREAT NIGHTINGALE †. It is certain that of nightingales there are varieties; but naturaliſts are much divided in regard to the haunts of the large kind of theſe birds, whether they frequent the plains, or the margin of waters. Schwenckfeld aſſerts that the ſmall ones ſettle on the pleaſant knolls: Aldrovandus ſays, that they live in the woods: others, on the con-

* Ariſtotle, *Hiſt. Anim.* lib. ix. 15. We muſt obſerve that, according to the Greeks, who are here the original authors, it was Progne that was metamorphoſed into a nightingale, and Philomela, her ſiſter, into a ſwallow. The Latin writers tranſpoſed or confounded the names, which has given currency to the error.

† Luſcinia Major, *Briſſ.*
Slowick Wickſzy, Rzacynſki.
Groſſe Nachtigalle, *Schwenckf.*
Sproſs-Vogel, *Friſch.*

trary,

trary, maintain with more probability, that ſuch as inhabit the dry foreſts which exclude the rain and the dew, are the ſmalleſt ſort. In Anjou, there is a kind of nightingales which are much larger than ordinary, and which lodge and neſtle among the hornbeams; and the little ones prefer the ſides of rivulets and pools. Friſch ſpeaks alſo of a breed ſomewhat larger than the common ſort, and which ſing more in the night, and in a manner rather different. Laſtly, the author of the treatiſe on the nightingale admits three kinds; the largeſt, ſtrongeſt, and beſt ſingers lodge among the buſhes near water; the middle-ſized live in the plains; and the ſmalleſt dwell in the mountains. From all this it appears that there is one or more breeds of large nightingales, but they are attached to a very permanent abode. The large nightingale is the moſt common in Sileſia; its plumage is cinereous with a mixture of rufous, and it is reckoned a better ſinger than the ſmall kind.

II. The White Nightingale *. This variety was very rare at Rome. Pliny relates, that a preſent was made of one to Agrippina, wife of the Emperor Claudius, which coſt 6000 ſeſterces †; valued by Budé at 15,000 crowns

* Luſcinia Candida, *Briſſ.*

† Pliny, *Hiſt. Nat.* lib. x. 29.

in

in his time, and which muſt be equal to double that ſum at preſent. Aldrovandus however aſſerts, that there is ſome miſtake in the figures, and that the amount is ſtill greater. That author ſaw a white nightingale, but mentions no particulars with regard to it. The Marquis d'Argence has one of this colour at preſent, which though young is very large, and its warble formed, but not ſo powerful as that of the old ones. "Its head and neck," ſays the Marquis, "are of the fineſt white; its wings and tail of the ſame colour: the feathers on the middie of the back are of a very light brown, and intermixed with ſmall white feathers . . . thoſe under the belly are of a white gray. This ſtranger ſeemed to cauſe wonderful jealouſy in an old nightingale which I have had for ſome time."

FOREIGN

FOREIGN BIRD

RELATED TO THE NIGHTINGALE.

THE FOUDI-JALA.

Motacilla Madagafcarienfis, *Gmel.*
Sylvia Madagafcarienfis, *Lath. Ind.*
Lufcinia Madagafcarienfis, *Briff.*
The Madagafcar Warbler, *Lath. Syn.*

THIS nightingale, which is found in Madagafcar, is of the fame fize with ours, and refembles it in many refpects; only its legs and wings are fhorter, and the colours of its plumage different. Its head is rufous, with a brown fpot on each fide; its throat white; its breaft light rufous; its belly brown, tinged with rufous and olive; all the upper part of its body, including what appears of the wing and tail quills, of an olive brown; its bill, and feet, deep brown. Briffon, who firft defcribed this fpecies, mentions not whether it fings; unlefs we fuppofe that the name of nightingale is alone fufficient to mark that faculty.

Total length, fix inches five lines; the bill, nine lines; the *tarfus*, nine lines and a half; the

the alar extent, eight inches and a half; the tail, two inches and a half, composed of twelve quills, somewhat tapering, and exceeding the wings by twenty lines. [A]

[A] Specific character of the *Motacilla Madagascariensis*: "It is greenish, below whitish; its throat and anus yellow; its eye-lids white."

THE

FAUVETTE.

FIRST SPECIES.

Motacilla-Hippolais, *Linn.*
Curruca, *Briss.*
The Pettychaps, *Lath.*

IN the dreary ſeaſon of winter, nature ſinks into inactivty and torpor. The inſects no more appear; the reptiles ceaſe to crawl; the vegetables are ſtripped of their verdure, and arreſted in their growth; the inhabitants of the air have periſhed, or retired to happier climes; thoſe of the waters are confined to their icy priſons, and moſt of the land animals have repaired to their caves and ſubterraneous manſions; the whole is a boundleſs picture of deſolation, and languor, and death. The vernal appearance of the feathered viſiters of the grove proclaims the return of animation and of joy. Nature awakens from her lethargy, and reſumes her enlivening powers. The trees ſpread their leafy ſhade; the vegetable tribes glow with their richeſt tints; the ſcene ſmiles around, it

warbles

No. 121

THE PETTYCHAPS.

warbles on every ſpray, and a genial fluſh heightens the whole.

Of the joyous gueſts of the woods, the fauvettes are the moſt numerous and the moſt charming: lively and volatile, each motion is expreſſive, each accent is cheerful, and each geſture diſplays the tenderneſs of love. Theſe pretty warblers arrive when the trees put forth their leaves, and begin to expand their bloſſoms; they diſperſe through the whole extent of our plains: ſome inhabit our gardens, others prefer the clumps and avenues; ſome conceal themſelves among the reeds, and many retire to the heart of large woods. Thus the fauvettes are diſperſed through every ſpot, and their ſprightly movements, and their cheerful airs, enchant each ſcene *.

Poſſeſſed of every grace and elegance, beauty alone was wanted to perfection. But nature has here checked her profuſion, and withheld decoration. Their plumage is dull and obſcure, except in two or three ſpecies, which are ſlightly ſpotted: all the reſt are ſtained with dark ſhades of whitiſh, gray, and ruſty.

The firſt ſpecies, which is the pettychaps,

* "In ſummer, a perſon cannot walk into any ſhady ſpot beſide brooks, but he will hear the fauvettes warbling even at the diſtance of a quarter of a league: this bird is known therefore in all countries." *Belon.*

is of the ſize of the nightingale. The robe of the nightingale is brown rufous, that of the pettychaps is brown-gray: it is alſo lightly tinged with ruſty gray, on the fringe of the coverts of the wings, and along the webs of the ſmall quills: the large quills are of a blackiſh cinereous; and ſo are alſo the quills of the tail, of which the two outermoſt are white on their exterior ſide, and at the tips on both ſides: over the eye there extends, from the bill, a white line like an eyebrow; and under the eye, and a little behind, there is a blackiſh ſpot; and this ſpot borders with the white on the throat, which has a ruſty caſt on the ſides, of a deeper colour under the belly.

This is the largeſt of all the fauvettes, except the Alpine warbler, of which we ſhall treat in the ſequel. Its total length is ſix inches; its alar extent, eight inches ten lines; its bill, from the tip to the angles, eight lines and a half; its tail, two inches ſix lines; its foot, ten lines.

It lives with other ſmaller ſpecies in gardens, in thickets, and in fields of peaſe and beans: they all ſit on the ſtakes which prop theſe; and there they frolic, build their neſt, and continually go out and in, till harveſt, which precedes their departure, forces them from this aſylum, or rather this abode of love.

It is amuſing to ſce them play together, grow angry, and chaſe each other: their attacks are gentle,

gentle, and their quarrels commonly end in ſongs. The pettychaps was regarded as the emblem of volatile amours, as the turtle was the image of ardent and ſteady love; yet the pettychaps, though lively and joyous, maintains a faithful and ſtrong attachment, while the turtle, all ſad and plaintive, infringes ſhamefully on the connubial rights. The male of the pettychaps laviſhes a thouſand little offices on his mate during incubation; he ſhares with her the ſolicitude for their tender young, and does not abandon her after the education of their family: his attachment outlives the appetite of fruition.

The neſt is compoſed of dry herbs and ſtalks of hemp, and lined with a little hair. It contains commonly five eggs, which the mother deſerts if they are touched: this approach of an enemy ſhe regards as a bad omen for the expected family. Nor is it poſſible to deceive her with the eggs of other birds. "I have often ſet ſtrange eggs under ſeveral ſmall birds," ſays Viſcount Querhoënt; "thoſe of the titmice under wrens, thoſe of the linnet under the redbreaſt: but I could never make the fauvettes hatch them; they always broke them; and, when I ſubſtituted other young ones, they ſoon killed them." From what wonderful inſtinct, if we believe the bulk of bird-catchers, and even of obſervers, do the pettychaps hatch the egg which the cuckoo drops into their neſt after

 deſtroy-

deſtroying their own; and how can they foſter the little ugly uſurper as their own progeny? However, it is the babbling fauvette or warbler to which this truſt is the ofteneſt committed, and perhaps that ſpecies has different inſtincts. —The pettychaps is of a timid character; it flies from birds as ſmall as itſelf, and with better reaſon it makes a rapid retreat from the ſhrike, its formidable enemy: but the danger is forgotten the moment it is paſt, and the little warbler again reſumes its cheerfulneſs, and friſks, and ſings. It is heard in the thickeſt boughs, where it is generally concealed; at times it pops out to the edge of a buſh, but hides itſelf again in an inſtant, particularly in the heat of the day. In the morning it gathers the dew; and, after the fall of a ſummer's ſhower, it trips among the wet leaves, and bruſhes off the drops.

Almoſt all the fauvettes retire at the ſame time, in the middle of autumn, and ſcarcely any remain in October. Their departure precedes the cold weather, which deſtroys the inſects, and blaſts the ſmall fruits, on which they ſubſiſt; for they not only catch flies, and gnats, and ſearch for worms, but they eat the berries of ivy, mezereon, and brambles: they grow fat during the maturity of the elders, dwarf elders, and privets.

The bill is ſlightly ſcalloped near the point: the

the tongue is fringed at the tip, and appears forked: the inſide of the bill is yellow at the bottom, and black towards the extremity: the gizzard is muſcular, and preceded by a dilatation of the *æſophagus:* the inteſtines are ſeven inches and a half long: there is generally no gall bladder, but two ſmall *cæca:* the outer toe is connected to the middle one by the firſt *phalanx*, and the outer nail is the ſtrongeſt of all. The teſticles in a male caught the 18th of June were five inches lengthwiſe, and the ſmaller diameter four inches. A female was diſſected on the fourth of the ſame month, and the oviduct was much dilated, and contained an egg, and the *ovarium* preſented a cluſter of unequal ſizes.

In the ſouthern provinces of France, and in Italy, moſt of the fauvettes are called epicurean warblers *(bec-figues)*; an error to which the nomenclators with their generic term *ficedula* have not a little contributed. Aldrovandus gives a confuſed and incomplete account of the ſpecies comprehended; and he ſeems not ſufficiently acquainted with them. Friſch remarks, that the genus of the fauvettes is the moſt obſcure and indetermined in the whole of ornithology. We have endeavoured to throw on it ſome light, by following the order of nature. All our deſcriptions, except that of a ſingle ſpecies, have been drawn from life; and it is from our own obſervations, and from the facts communicated by intelligent obſervers, that we have delineated the diſtinctions and the ſimilitudes, and the habits which obtain among theſe little birds. [A]

[A] Nothing can exceed the confuſion which nomenclators have introduced into the article of the pettychaps, or fauvette. Gmelin and Latham have transferred the Greek name, *hippolais*, which Linnæus had injudiciouſly applied to that bird, and have beſtowed it on another bird about one third of the ſize; and at the ſame time they have given the pettychaps, or fauvette, the epithet *hortenſis*. Yet while theſe two authors agree in the application of the terms, the one aſſerts that the *motacilla hortenſis* is larger than the redpole or black-cap, but the other repreſents it as ſmaller than even the linnet. It will be unneceſſary therefore to tranſlate the ſpecific characters. The *motacilla-hippolais* of Gmelin, or the leſſer pettychaps of Latham, is ſaid to build in the hedges near the ground; its egg white, ſprinkled with numerous minute red ſpecks.

THE

PASSERINETTE, OR LITTLE FAUVETTE, *Buff.*

SECOND SPECIES.

Motacilla Pafferina, *Gmel.*
Sylvia Pafferina, *Lath. Ind.*
Curruca Minor, *Briff.*
Mufcicapa Secunda, *Aldrov. Ray.* & *Will.*
The Pafferine Warbler, *Lath. Syn.*

WE adopt the name *pafferinette*, which this bird receives in Provence. This is a fmall fauvette, and is diftinguifhed from the preceding, not only by its fize, but by its plumage, and by the monotonous burthen *tip*, *tip*, of its fhort fong, which it continually repeats as it hops among the bufhes. A very delicate white gray covers all the fore and under part of the body, receiving a very light brown caft on the fides: an uniform afh gray is fpread over the whole of the upper part, and ftained fomewhat with blackifh on the great quills of the wings and of the tail: there is a fmall whitifh ftreak which paffes over the eye. Its

 length

length is five inches three lines; and its alar extent eight inches.

The passerinette makes its nest near the ground, among shrubs: we saw one in a goose-berry bush in a garden; it was like a half-cup, composed of dry herbs, rough on the outside, but finer and better interwoven within: it contained four eggs, of a dirty white ground, with green and greenish spots, spread thicker near the large end. The iris is chesnut, and there is a very small scalloping near the point of the upper mandible: the hind nail is the strongest: the feet are lead-coloured: the intestinal tube from the gizzard to the anus is seven inches, and there are two inches from the gizzard to the *pharynx*: the gizzard is muscular, and preceded by a dilatation of the *œsophagus*: no gall bladder could be found, nor *cæcum*.—The subject was a female: the rudiments of the eggs in the ovarium were of unequal sizes. [A]

[A] Specific character of the *Motacilla Passerina*: "It is cinereous, the under side white gray; the eyelids whitish, the wing quills and tail black." This bird is unknown in England.

N.o 122

FIG 1. THE BLACK CAP. FIG. 2. THE EPICUREAN WARBLER.

THE

BLACK-HEADED FAUVETTE, *Buff.*

THIRD SPECIES.

Motacilla Atricapilla, *Linn. Gmel. Scop. Brun. Kram. &c.*
Sylvia Atricapilla, *Lath. Ind.*
Curruca Atricapilla, *Briſſ. & Klein.*
Atricapilla, ſeu Ficedula, *Geſn. & Aldr.*
The Black-cap *, *Penn. Will. & Lath. Synop.*

ARISTOTLE, enumerating the various changes which the revolution of the ſeaſons produces on the feathered tribes, ſays that the beccafico or epicurean warbler is metamorphoſed in autumn into the black-cap †. Naturaliſts have been much puzzled with this aſſertion; ſome regard it as marvellous, others reject it as incredible ‡: but it is really neither the

* In Greek, Μελανηορυφος, Μελανηκεφαλος: in Italian, *Capinera, Capinegro:* in German, *Graſz-muckl, Graſz Spatz:* in Saxon, *Monch, Monchlein:* in Swiſs, *Schwartz-Kopff:* in Bohemian, *Plaſk:* in Poliſh, *Figoiadka.*

† *Hiſt. Anim.* lib. ix. 49.

‡ *Niphus*, in Aldrovandus, ſtrains at a ſolution of the problem, by diſtinguiſhing a great and little *black-head*; the latter not being tranſmuted into a beccafico, but the other being never ſeen at the ſame time, and actually undergoing the

the one nor the other; and the explication is very eafy. In fact, the young black-caps have, through the whole fummer, the plumage of the epicurean warblers, and only affume their proper garb after the firft moulting: and this is the interpretation which Pliny gives*.

Aldrovandus, Johnfton, and Frifch, after defcribing the black-cap, introduce a fecond fpecies, which has a brown head†: but this is only the female of the former, and the fole difference of appearance between the two fexes confifts in the colour of the head. In the male, a black cap covers the back of the head and the crown, as far as the eyes; below and round the neck the plumage is of a flate gray, lighter on the throat, attenuated into white on the breaft, and fhaded with blackifh on the fides; the back is of a brown gray, lighter on the exterior furface of the quills, deeper on the lower ones, and ftained with an olive tint. The bird is five inches five lines in length; the alar extent eight inches and a half.

the metamorphofis. "The Bolognefe bird-catchers," fays Aldrovandus, "thus diftinguifh them;" yet he will not admit that opinion, and the moment after he confounds the black-cap with the bulfinch.

* *Hift. Nat.* lib. x. 44.

† Atricapilla altera, *Johnft.*
Atricapilla alia caftaneo vertice, *Aldrov.*
Curruca vertice fubrubro, *Frifch.*

The

The black-cap has the moſt pleaſant and the fulleſt warble of all the fauvettes. It is ſomewhat like the nightingale's ſong, and we enjoy it much longer; for ſeveral months after the groves no more echo Philomel's notes, the muſic of the black-cap is heard. Its airs are eaſy and light, and conſiſt of a ſucceſſion of modulations of ſmall compaſs, but ſweet, flexible, and blended: they expreſs the happineſs and tranquillity that dwell in their haunts. The ſenſible heart warms with delicious emotions at accents inſpired by nature, and flowing from that felicity which ſhe has beſtowed.

The male ſhews a tender concern for his female: not only does he carry flies, worms, and ants to her; but he relieves the languor of incubation, and ſits by turns. The neſt is placed near the ground, and carefully concealed in a coppice: it contains four or five eggs, of a greeniſh hue, with ſpots of light brown. The young ones grow in a few days; and though but ſlightly fledged, they will leap out of the neſt when a perſon comes near it, and never will return. The black-cap has generally only one annual hatch in France. Olina ſays that it makes two in Italy; and ſuch muſt be the caſe with many other kinds of birds which inhabit a warmer climate, where the ſeaſon of love is prolonged.

At

At its arrival in the ſpring, if the inſects are deſtroyed by the relapſe of cold, the black-cap has recourſe for ſubſiſtence to the berries of ſome ſhrubs, as thoſe of the ſpurge laurel and ivy; in autumn they alſo eat the ſmall ſeeds of the berry-bearing alder, and of the hunters ſervice-tree *. During that ſeaſon they often go to drink, and about the end of Auguſt they are caught near the ſprings: they are then exceeding fat, and of a delicate taſte.

The black-cap may be alſo raiſed in the cage; and of all the birds of the volary it is, ſays Olina, the moſt lovely †. The attachment which it ſhews to its maſter is charming; it welcomes him with a peculiar accent, and a more tender air. On his approach it darts towards him againſt the wires of the cage, and ſtruggles to burſt its priſon to meet him; and by the continual flapping of its wings, with its feeble cries, it ſeems to expreſs its tranſports of joy ‡.

The young ones bred in a cage, if they be within hearing of the nightingale, will improve

* Schwenckfeld.

† "Beyond the other birds of the cage, it is of a cheerful diſpoſition, with a ſweet and delightful ſong, with a lovely and pleaſing aſpect." *Olina, Uccelleria*, p. 9.

‡ Olina, p. 9. Of this bird Mademoiſelle Deſcartes ſaid, "No offence to my uncle, it has ſentiment."

their

their ſong, and rival their maſter*. In the ſeaſon of their departure, which is the end of September, all theſe priſoners are reſtleſs and uneaſy in their confinement, particularly during the night and while the moon ſhines. They ſeem conſcious of the migration which they ſhould now perform; and ſo ardent is their deſire of changing their climate, that at this time many die from vexation and diſappointment.

This bird is common in Italy, France, Germany, and even in Sweden; yet it is ſaid to be unfrequent in England †.

Aldrovandus ſpeaks of a variety of this ſpecies which he calls the *variegated beccafico*, or fig-pecker *(ficedula)*; but he does not inform us whether it is only an individual or a permanent difference. Briſſon, who mentions it under the appellation of *black and white fauvette*, gives no further notice; and it would ſeem that the *black-backed fauvette* of Friſch is only the ſame variety.

* "The black-cap which I raiſed has formed its ſong after the nightingale, and has extended its voice to ſuch degree, as to ſilence its maſters, my nightingales."

Note communicated by M. le Treſorier le Moine.

"The young ones caught with the net will perfect their ſylvan ſong, and adopt other ſorts of airs from tame linnets or other birds, and will teach their neſtlings all that they have acquired." *Olina.*

† Willughby.

 The

The *little pigeoner (petite colombaude)* of the Provencals is another variety of the black-cap; only it is rather larger, and all the upper part of its body is of a deeper colour, almoſt blackiſh; its throat is white, and its ſides gray: it is neat and ſprightly; is fond of ſhades, and of the cloſeſt woods, and delights in the dew, which it eagerly collects.

In a hen black-cap opened the fourth of June, the *ovarium* contained eggs of various ſizes; the inteſtinal tube from the *anus* to the gizzard was ſeven inches and a quarter long; there were two diſtinctly formed *cæca*, two lines in length: the tongue was ſlender, and forked at the end; the upper mandible ſlightly ſcalloped; the outer toe joined to the middle one by its firſt *phalanx*; the hind nail the longeſt of all.

In a cock which was diſſected on the 19th of June, the teſticles were four lines long, and three broad: the *trachea arteria* had a knot ſwelled where it forks; the *œſophagus* about two inches long, and formed a ſac before its inſertion into the gizzard.

[A] Specific character of the Black-cap, *Motacilla-Atricapilla*: "It is brick-coloured, below cinereous, with a dark cap." "The black-cap," ſays Mr. White, "has a full, ſweet, deep, loud, and wild pipe; yet that ſtrain is of ſhort continuance, and his motions are deſultory: but when that bird ſits calmly and engages in ſong in earneſt, he pours forth very ſweet, but inward melody, and expreſſes great variety of ſoft and gentle modulations, ſuperior perhaps to thoſe of any of our

our warblers, the nightingale excepted. Black-caps moftly haunt our orchards and gardens: while they warble, their throats are wonderfully diftended." In Norfolk they are called the *mock nightingale*. Their egg is reddifh brown, with dufkier clouds, with ftraggling blackifh fpots.

THE

GRISETTE OR THE GRAY FAUVETTE,

Called in Provence, PASSERINE, *Buff.*

FOURTH SPECIES.

Motacilla Pafferina, *Gmel.*
Sylvia Pafferina, *Lath. Ind.*
Stoparola, *Aldrovandus.*
Curruca Minor, *Briff.*
The Pafferine Warbler, *Lath. Synop.*

ALDROVANDUS fpeaks of this bird under the name of *Stoparola*, which was given by the fowlers of Bologna, probably, fays this naturalift, becaufe it frequents the bufhes and thickets where it builds its neft *.

We have feen one of thefe nefts in a black thorn three feet from the ground; it was of a

* From the Italian *Stoppia*, ftubble or brufhwood.

cup

cup ſhape, and conſiſted of meadow moſs interwoven with a few ſtalks of dry herbs.—Sometimes it is formed entirely with theſe ſtalks, which are finer in the inſide, and coarſer on the outſide. The neſt contained five eggs of a greeniſh gray, ſprinkled with ruſty and brown ſpots, which are more frequent at the obtuſe end.

The mother was caught with her young: the iris was of a cheſnut colour; the edges of the upper mandible lightly ſcalloped at the point; the two eyelids furniſhed with white laſhes: the tongue was frittered at the end; the inteſtinal tube from the gizzard to the *anus* was ſix inches long: there were two *cæca* two lines in length, their diſtance two inches, and the firſt before its inſertion made a dilatation: the *ovarium* contained different ſized eggs.

In a male which was opened in the middle of May, the bowels preſented very nearly the ſame appearances: there were two teſticles, of which the right one was larger than the left, its great diameter four lines, and its ſmall diameter two lines and three quarters: the gizzard was muſcular, and the two membranes were detached; it contained ſome fragments of inſects, but no pebbles: the iris was light crimſon; in another it appeared orange; which ſhews that this part is liable to vary in its colours, and cannot furniſh a ſpecific character.

Aldrovandus

Aldrovandus remarks, that the eye of the passerine is small, but brisk and lively. The back and crown of the head are ash gray: the temples, the plumage above and behind the eye, are marked with a more blackish spot: the throat is white as far as the eye: the breast and stomach are whitish, and shaded with a light rusty or vinous tint. The bird is larger than the epicurean warbler: its total length is five inches seven lines; its alar extent eight inches. In Provence it enjoys another climate, and its habits are rather different. It likes to repose under the fig-tree and the olive, feeds on their fruits, and its flesh becomes extremely delicate. Its feeble notes seem to repeat the two last syllables of its name, *passerine*.

M. Guys sent us from Provence a small kind of fauvette, under the name of *bouscarle*, engraved *Pl. Enl.* No. 655. fig. 2. It seems to be most related to the gray fauvette, or passerine warbler; but its colour is rather fulvous and brown than gray. [A]

[A] Specific character of the *Motacilla Passerina*: "It is cinereous, below gray white; its eyebrows whitish, its wing quills and tail black."

THE

THE

BABBLER FAUVETTE, *Buff.*

FIFTH SPECIES.

Motacilla-Curruca, *Linn. Gmel. Scop. Mull. Frif.*
Sylvia-Curruca, *Lath. Ind.*
Curruca Garrula, *Briff. & Klein.*
Ficedula Canabina, *Will.*
The Babbling Warbler, *Lath. Syn.**

WE hear this warbler the ofteneft, and almoft continually in fpring. It frequently mounts a fmall height directly over the hedges, and whirls in the air and drops back again, chanting a fhort paffage of a lively joyous air, which is always the fame, and which it inceffantly repeats: hence it has received the epithet of *babbler*. Befides this burthen, which it fings ofteneft while on the wing, it has another found or hollow whiftle, *bjie, bjie*, which it makes in the heart of the bufhes, and which we could hardly imagine to be uttered by fo

*In Greek Υπολαις, Επιλαις: in modern Greek, Πολαμιδα: in Italian, *Pizamofche, Becafico Canapino*: in German, *Grafsmuck, Fable Gras-muck*: in Polifh, *Piegza*: in Swedifh, *Kruka*.

little

little a bird. Its motions are as ſprightly and frequent as its babble is conſtant; and it is the moſt friſky and alert of all the fauvettes. It is perpetually buſtling, fluttering, hopping in and out among the buſhes, without allowing a moment's reſt. It neſtles in the hedges, along the high roads, in the ſpots which afford it ſhelter, and commonly near the ground, and on the tufts of graſs which ſpring up among the roots of the buſhes *: its eggs are greeniſh dotted with brown.

According to Belon, the modern Greeks call this fauvette *potamida*, *i. e.* bird of rivers or rivulets. Such is the name it has in Crete; and perhaps in a warm climate † it affects the neighbourhood of waters more than in our temperate countries, where it can eaſily procure cooling moiſture. The inſects bred by heat and moiſture conſtitute its chief food. The name which Ariſtotle gives it ‡ implies that it continually ſearches for worms; yet it is ſeldom ſeen on the ground,

* Schwenckfeld.

† Belon, p. 340.—"There is another bird called by the ancients *curruca*, which the French know under the name of *brown fauvette*, and which the Greeks who at preſent inhabit this iſland (Crete) call *potamida*. They hold that the cuckoo is hoſtile to it, and eats the young when it has an opportunity." *Dapper*, *Deſcrip. des Iles de l'Archipel.*

‡ Υπολαις, which Geſner tranſlates *Curruca*. From ὑπο and λαος, a ſtone; becauſe it gropes under ſtones for worms.

and the reptiles which it feeds on are the caterpillars it finds on the ſhrubs and buſhes.

Belon at firſt calls it the *brown fauvette*, and afterwards he beſtows the epithet of *leaden*, which marks much better the real tint of its plumage. The crown of its head is cinereous; all its robe aſh brown; the fore part of its body white ſtained with ruſty; the wing quills brown, their inner edge whitiſh: the outer edge of the great quills is cinereous, and that of the middle ones ruſty gray: the twelve quills of the tail are brown edged with gray, except the two outermoſt, which are white on the outſide, as in the common fauvette or pettychaps: the bill and feet are leaden gray: it is five inches long, and its alar extent ſix inches: it is of the ſame ſize with the griſette or paſſerine warbler, and on the whole reſembles it much.

To this ſpecies we muſt refer not only the *hemp-beccafico* of Olina, which he ſays is frequent among the hemp-fields of Lombardy, but alſo the *canevarola* of Aldrovandus, and the *titling* of Turner.—This bird is eaſily tamed: as it lives in our meadows, our thickets, and our gardens, it is already half domeſticated. If it is to be bred for the cage, which is ſometimes done for the ſake of its cheerful ſong, we muſt, ſays Olina, wait till it be fledged, and then take it from the neſt, and put a bathing-cup in the cage;

cage; for, without this precaution, it would die. And with proper care its life may be prolonged to eighteen years in confinement. [A]

[A] Specific character of the *Motacilla-Curruca*: "Above it is brown, below whitish; its tail quills brown; the outermost with a narrow white edging." It inhabits from Italy to Siberia. Its egg is cinereous, with rusty spots.

THE

RUSSET, OR FAUVETTE OF THE WOODS, *Buff.*

SIXTH SPECIES.

Motacilla-Schœnobænus, *Linn. & Gmel.*
Sylvia Schœnobænus, *Lath. Ind.*
Curruca Sylvestris, seu Lusciniola, *Briss. Ray, & Will.*
The Bog-rush Warbler, *Penn.*
The Reed Warbler, *Lath. Syn.*

IF Belon had not expressly distinguished the *russet* or *fauvette of the woods* from his *mouchet*, which we shall find to be the *winter fauvette* or hedge sparrow, we should have considered these as constituting the same species. Nor are we convinced that they are different

 birds,

birds, ſince their reſemblance is ſo great, and their diſcrimination ſo little: we only yield to the authority of Belon, who has perhaps obſerved them better than we have done.

Like the reſt of the fauvettes, this bird is perpetually joyous, lively, and active, and often utters a feeble cry: it has alſo a ſong, which though monotonous is not diſagreeable; and it improves the notes when it has opportunities of hearing more varied and more brilliant modulations *. Its migrations ſeem not to extend beyond our ſouthern provinces; there it appears in winter †, and ſings in that ſeaſon: in ſpring, it returns to our woods, preferring the copſes, and builds its neſt with green moſs and wool: it lays four or five eggs, which are a ſky blue.

The young ones are eaſily raiſed and bred, and they amply repay the trouble of education by their familiarity, their pretty warble, and their cheerfulneſs. Nor are they deſtitute of courage. "Thoſe which I trained," ſays De Querhoënt, "were the terror of many birds as large as themſelves. In the month of

* "Thoſe which I raiſed ſeemed to have a more melodious ſong than the wild ones, becauſe they pretty often heard a fiddle. They ſang frequently." *Note de M. le Vicomte de Querhoënt.*

† "It does not leave the country, and ſings in winter like the gold-creſted wren." *Id.*

April

April I ſet all my little priſoners at liberty; but the ruſſets were the laſt to profit by it. As they often made ſhort excurſions, the wild birds of the ſame ſpecies purſued them: but they ſheltered themſelves on the ſole of my window, where they ſtoutly defended their poſt: they briſtled their feathers; each party trilled a feeble ſtrain, and pecked the board like cocks, and ſo entered into a keen combat."

This is the only fauvette which we have not been able to delineate from nature.—The deſcription which is given of its plumage confirms us in the opinion, that this ſpecies is at leaſt much related to the hedge ſparrow, if not exactly the ſame. Its head, the upper ſurface of its neck, the breaſt, the back, and the rump, are variegated with brown and rufous, each feather being brown in the middle, and edged with rufous; the ſcapular feathers, the coverts of the upper part of the wings and of the tail, variegated with the ſame colours, and in the ſame manner; the throat, the lower part of the neck, the belly, and the ſides, ruſty; the quill feathers of the wings brown, and edged with rufous; thoſe of the tail entirely brown. It is of the ſize of the pettychaps. The plumage of the fauvettes is in general dull and obſcure; that of the ruſſet is one of the moſt variegated, and

Belon deſcribes with warmth the beauty of its colours *. He remarks, at the ſame time, that this bird is ſcarcely known except to the fowlers and the peaſants who live near the woods †, and that it is caught in the heats of the ſummer, when it drinks at the pools. [A]

[A] Specific character of the *Motacilla-Schœnobænus*: "It is brown brick-coloured, below pale brick; its head ſpotted."

THE

REED FAUVETTE.

SEVENTH SPECIES.

Motacilla Salicaria, *Linn.* & *Gmel.*
Sylvia Salicaria, *Lath. Ind.*
Curruca Arundinacea, *Briſſ.*
Luſcinia Salicaria, *Geſner*, *Ray*, *Will.* & *Klein.*
Avis Stoparolæ ſimilis, *Sibbald.*
The Willow Lark, *Penn.*
The Sedge Warbler, *White*, *Albin*, & *Lath. Syn.*‡

THE reed fauvette chants in the warm nights of ſpring like the nightingale, which has occaſioned ſome to call it the willow or oſier

* *Nat. des Oiſeaux*, p. 338. † *Idem.*

‡ In German, *Weiderich*, *Wydenguckerlin*: in Swiſs, *Weiderle*, *Zilzepſle*: in Poliſh, *Bowniоſka*.

oſier nightingale. It makes its neſt among reeds and buſhes, amidſt marſhes, and in copſes beſide the margin of pools. We ſaw one in the low branches of a hornbeam, near the ground; it conſiſted of ſtraw and ſtalks of dry herbs, with a little hair within. It is conſtructed with more art than that of the other fauvettes, and uſually contains five eggs of a dirty white, mottled with brown, which is deeper and more ſpread about the thick end.

The young ones, though tender and not fledged, deſert the neſt if it be touched, or even if a perſon go too near it: this feature, which is common to all the fauvettes, and even to this ſpecies which breeds amidſt water, ſeems to characterize the inſtinctive diſpoſition of theſe birds.

During the whole of the ſummer we ſee it darting from among the reeds, to catch the dragon-flies, and other inſects which buz on the ſurface of the water. It continually warbles*; and it drives away the other birds†, that it may remain ſole proprietor of its ſpot, which it does not quit till September, the ſeaſon when it departs with its family.

It is of the ſize of the black-cap; being five inches and four lines in length, and its alar ex-

* Hebert. † Geſner.

tent eight inches eight lines: its bill is ſeven inches and a half long; its feet, nine lines; its tail, two inches: the wings, when cloſed, reach beyond the middle of the tail: all the upper part of its body is of a light ruſty gray, and inclining ſomewhat to olive near the rump: the feathers of the wings are browner than thoſe of the tail: the inferior coverts of the wings are of a light yellow; the throat and all the fore part of the body yellowiſh on a whitiſh ground, and ſtained on the ſides and near the tail with brown ſhades.

It is not in the leaſt degree probable that the *petronella* of Schwenckfeld, "a bird which neſtles under rocks and on the bare ground, which is ſeen only in the craggy parts of the mountains, and which continually jerks its tail like the wagtail,": is the ſame with our reed fauvette. We cannot conceive why Briſſon ranged them together; for even the plumage which Schwenckfeld deſcribes, would ſhew it to be rather a kind of redſtart.

If the *ſedge bird* of Albin is alſo the ſame, his figure muſt be a very bad one, and all its colours falſe: it is not painting but maſking nature. The figure given by Aldrovandus, and borrowed from Geſner, under the name of *ſalicaria*, has a much thicker bill than belongs to the genus of fauvettes; and if the bird *(avis*

con-

consimilis stoparolæ & magnanimæ) is the reed warbler, as Brisson says, and which seems probable, it will be difficult to suppose that the *salicaria* is the same. Such is the confusion of Aldrovandus's account of this genus, which he seems not to have known from his own observations; and the example of this respectable naturalist shews how dangerous it is to trust to defective or inaccurate relations. [A]

[A] Specific character of the Sedge Warbler, *Motacilla Salicaria*: "It is cinereous, below white; its eye-brows white." It is not uncommon in England; sings night and day in the breeding season, imitating the notes of a sparrow, of a swallow, and of a sky-lark.

THE

LITTLE RUFOUS FAUVETTE.

EIGHTH SPECIES.

Motacilla Rufa, *Gmel.*
Sylvia Rufa, *Lath. Ind.*
Curruca Rufa, *Briss.*
Múscipeta Minima, *Fris.*
The Rufous Warbler, *Lath. Syn.*

BELON tells us, that he was at great pains to discover the ancient name of the little rufous fauvette, and yet in settling this point he he falls into a miſtake, conceiving it to be the *troglodyte.* He ſeems even ſenſible, in ſome meaſure, of his error; for he obſerves that the text of *Ætius* and *Paul Æginetus,* which deſcribes the *troglodyte,* agrees better with the brown wren than with the rufous fauvette. And we ſhall afterwards find that this remark is well founded. Indeed the appellation of *troglodyte* can refer only to a bird which frequents caverns, and the holes of rocks or of walls, a character which belongs to none of the fauvettes; though Belon, erroneouſly imagining the word *fauvette* derived from the Latin *fovea,* a pit or burrow, admits it to have this inſtinct *.

The rufous warbler has commonly five young; but they often become the prey of the rapacious birds, particularly the ſhrikes. The eggs are greeniſh white, and marked with two kinds of ſpots; ſome obſcure and hardly viſible, ſcattered equally over the ſurface; others deeper and well defined, moſt frequent near the thick end. "It conſtantly makes its neſt," ſays Belon, "in ſome garden herb or buſh, ſuch as hemlock and the like, or behind a garden wall in the towns or villages." The inſide is lined with horſe-hair; but the neſt obſerved by Belon had

* *Fauvette* is really derived from *fauve,* fox-colour. *Menage.*

a hole

a hole in the bottom, which he afcribes to defign, though it was probably accidental*; for this is contrary to the general conftruction, which is calculated to collect and concentrate the heat.

The fame naturalift hits better when he fays that the plumage of this little warbler is uniform, and the fame with that of the nightingale's tail. The comparifon is happy; and will fave us a minute defcription. We fhall only obferve, that there is a little rufous fhading the great coverts of the wings, and more faintly fpread through the webs of their quills, with a very dilute and light tinge of rufty on the gray of the back and head, and on the whitifh colour of the fides. This bird is therefore improperly ftyled *the rufous*, fince only a few parts of its plumage are dafhed flightly with it.

Its total length is only four inches eight lines; its alar extent fix inches ten lines: it is one of the fmalleft of the genus, being inferior even to the pafferine warbler. But Belon feems to exaggerate when he fays, "that it is hardly fo big as the end of the finger." [A]

[A] Specific character of the *Motacilla Rufa*: "It is gray rufous, below tawny; a longitudinal ftreak on its temples; the quills of its wings and tail tawny."

* It is lined on the infide with horfe-hair, and fo nicely that it is perforated like a noofe; fo that the excrements of the young efcape, and they are always preferved clean." *Nat. des Oif.* p. 341.

THE

THE

SPOTTED FAUVETTE, *Buff.*

NINTH SPECIES.

Motacilla Nævia, *Gmel.*
Curruca Nævia, *Briff.*
Sylvia Nævia, *Lath. Ind.*
Boarina, *Aldrov.*
The Fig-eater, *Alb. & Lath.*

THE plumage of the fauvettes is commonly uniform and unvaried. The prefent is diftinguifhed by fome black fpots on the breaft; but the reft of its plumage is fimilar to that of the reft of the genus. It is of the fize of the fecond fpecies, or the pafferine warbler; its length five inches four lines, and its wings when clofed cover half the tail: all its mantle from the crown of the head to the origin of the tail is variegated with rufty brown, yellowifh and cinereous: the quills of the wings are blackifh, edged exteriorly with white: thofe of the tail are the fame: the breaft is yellowifh, marked with black fpots: the throat, the fore part of the neck, the belly, and the fides are white.

This

This warbler is more common in Italy, and probably in the ſouthern provinces of France, than in the northern countries, where it is little known. According to Aldrovandus, it is frequent near Bologna; and the name which he gives to it, ſhews that it uſually follows the herds of cattle in the fields *.

It builds in the meadows, and places its neſt within a foot of the ground in ſome large plant, as fennel, chervil, &c. It never ſprings when one approaches the ſpot, and it ſuffers itſelf to be caught rather than abandon its young, preferring the life of its progeny to its own: ſo powerful that inſtinct which inſpires the feeble, fugacious animals with courage and intrepidity! In all creatures that obey the wiſe laws of nature, the parental affection is the ſource of whatever may be deemed virtuous.

* *Boaro*, in Italian, ſignifies a cow-herd.

THE

THE

WINTER FAUVETTE,

Or Traine-Buisson, or Mouchet, *Buff.*

TENTH SPECIES.

Motacilla Modularis, *Linn. Gmel. Mull. Fris.*
Sylvia Modularis, *Lath. Ind.*
Curruca Sepiaria, *Briss.*
Sylvia Gulâ Plumbeâ, *Klein.*
Curruca Eliotæ, *Ray. & Will.*
The Hedge-Sparrow, *Penn. Alb. Will. & Lath. Ind.**

ALL the other fauvettes depart in autumn; this, on the contrary, arrives in that season. It resides among us during the whole of the winter months; and hence it has been styled the *winter fauvette*, and in some provinces the *winter nightingale*. The English and Italian appellations of *hedge sparrow*, and *wood-sparrow (passara salvatica)*, allude to the resemblance which its plumage, variegated with black and rufous brown, bears to that of the tree-sparrow; a resemblance which Belon found to be complete †.—In fact, the colours of the winter fauvette

* In Italian, *Passara Salvatica*: in German, *Prunell*: in Svedish, *Jaern-Spart.*

† *Nat. des Oiseaux*, p. 375.

No. 123

THE HEDGE WARBLER.

vette are much deeper than thoſe of the others: its general complexion is blackiſh, and all its quills and feathers are bordered with rufous brown: its cheeks, its throat, the fore part of its neck and breaſt are of a blueiſh cinereous: there is a ruſty ſpot on the temple: the belly is white. Its ſize is that of the red-breaſt; its alar extent eight inches. The cock differs from the hen, in having more of the rufous caſt on the head and neck, and the latter being more ſtained with cinereous.

Theſe birds perform their migrations in bodies: they arrive in the end of October, and the beginning of November: they alight on the hedges, and go from buſh to buſh, always near the ground, and hence their name of *trail-buſh (traîne-buiſſon).* It is not timorous, and is eaſily enſnared *. It has neither the ſhyneſs nor the vivacity of the other fauvettes, and its diſpoſition ſeems to participate of the cold and torpor of the ſeaſon.

Its uſual ſtrain is quivering; it is a ſort of ſoft ſhake *tittit-tititit*, which it often repeats. It has alſo a ſlender warble, which, though mournful and little varied, is pleaſant to hear in a ſeaſon when all the other ſongſters are ſilent: this is the moſt frequent and lengthened towards evening. In the depth of winter, the hedge-ſparrow haunts

* Willughby.

the

the barns and threſhing-floors, to pick up the fine meal from among the chaff. Hence probably the name *chaff-ſcraper* (*gratte-paille*), which is given to it in Brie. Hebert ſays, that he found whole grains of wheat in its craw; but its ſlender bill is not calculated for ſuch food, and neceſſity alone can compel it to that reſource. As ſoon as the cold abates, it again retires to the hedges, ſearching on the branches for the chryſalids, and dead vine-fretters.

It diſappears in the ſpring; whether that it penetrates into the foreſts and returns to the mountains, as in Lorraine, where I am informed that it breeds; or whether it migrates into other climates, particularly towards the north, from whence it ſeems to come in the autumn, and where it is very frequent in ſummer. In England, according to Albin, it is found during the warm weather in every buſh. It inhabits Sweden; and the epithet which Linnæus applies, ſeems to ſhew that it continues during the winter, and aſſumes the white plumage common in the northern climates in that ſeaſon *. It alſo breeds in Germany; but its neſt is very rarely found in France: it is placed near the ground, or even on the ſurface, and it conſiſts of moſs, lined with wool and

* *Paſſer Canus*: Syſt. Nat. edit. vi.

hair: it usually contains four or five eggs of a pleasant uniform light blue, without any spots. When a cat, or any mischievous animal, happens to come near the nest, the mother will divert it from the spot by an instinct similar to that by which the partridge misleads the dog; she springs up, flutters from spot to spot, till her enemy is removed to a safe distance. Albin says, that in England the young are hatched against the month of May, that they are easily raised, that they are not timorous, and even become very familiar; and lastly, that their warble is esteemed, though not so cheerful as that of the other fauvettes *.

Their leaving France in the spring, and their plenty in the northern regions during that season, are singular facts in the history of the migration of birds. After the grasshopper warbler, this is the second species with a slender bill,

* A winter fauvette kept during that season at the house of M. Daubenton the younger, and caught in a snare in autumn, was not wilder than if it had been taken from the nest. It was put into a volery filled with canaries, linnets, and goldfinches. A canary took such a liking to this fauvette that he would never leave it; and M. Daubenton was induced to remove them from the general volery, and put them by themselves in a breeding-cage. But this attachment seemed to be friendship only, and not love; they did not copulate, nor is it likely that their union would have been productive.

which retires from the heats of our ſummers, and yet ſupports the rigours of our winters, which all the reſt of the genus ſhun: and this inſtinct alone is ſufficient to diſtinguiſh it, or at leaſt to ſet it at a ſmall diſtance from the others.

THE

ALPINE FAUVETTE.

THIS bird is found on the Alps and the high mountains of Dauphiné and Auvergne: it is at leaſt as large as the common bunting, and therefore in point of ſize it far exceeds the fauvettes; but ſtill it is connected to them by many marked characters. Its throat is white, ſpotted with two different tints of brown: its breaſt is aſh gray: all the reſt of its body is variegated with gray, more or leſs inclined to whitiſh, and with rufous: the inferior coverts of its tail are marked with blackiſh and white: the upper part of its head and neck is aſh gray: its back is of the ſame colour, but variegated with brown: the ſuperior coverts of its

No. 124

THE ALPINE WARBLER.

its wings are blackiſh, ſpotted with white to the point: the quills of its wings are brown, edged exteriorly, the large ones with whitiſh, the middle ones with ruſty colour: the ſuperior coverts of its tail are brown edged with greeniſh gray, and ruſty near the point: all the quills of its tail are terminated above by a ruſty ſpot on the inner ſide: its bill is eight lines in length, blackiſh above, yellow below at the baſe, and not ſcalloped: its feet are yellowiſh: the *tarſus* is an inch long: the hind nail is much thicker than the reſt: the tail is two inches and a half long, ſomewhat forked, and exceeds the wings near an inch. The whole length of the bird is ſeven inches: the tongue is forked; the *œſophagus* is rather more than three inches, and it dilates into a ſort of glandulous ſac before its inſertion into the gizzard, which is very thick, and an inch long, and eight lines broad: it is muſcular, and lined looſely by a membrane: it generally contains fragments of inſects, different ſmall ſeeds, and minute gravel. The left lobe of the liver, which covers the gizzard, is ſmaller than uſual in birds: there is no gall-bladder, but two *cæca* of a line and an half each: the inteſtinal tube is ten or twelve inches long.

Though theſe birds inhabit the Alpine tracts which lie between France and Italy, and even thoſe in Auvergne and Dauphiné, no author has mentioned them. The Marquis de Piolenc ſent ſeveral to M. Gueneau de Montbeillard, which were killed at his barony of Montbel, 18th January 1778. They never remove far from the lofty mountains, unleſs they be compelled to retreat by the abundance of ſnow: accordingly, they are hardly ſeen in the low country. They are generally on the ground, and run ſwiftly, ſcudding along like the quail and the partridge, and not hopping as the other fauvettes do. They alſo ſit upon ſtones, but ſeldom perch on trees: they wander in ſmall bodies, and recall each other by a feeble cry like that of the wagtail. When the cold is moderate, they live in the fields; but when it becomes more ſevere, they reſort to the moiſt meadows where there is moſs, and are then ſeen running on the ice. Their laſt reſource is the tepid ſprings and brooks: they are often found in ſuch ſituations when the perſon is hunting for ſnipes. They are not ſhy; yet are they difficult to kill, eſpecially on the wing.

THE

PITCHOU.

Motacilla Provincialis, *Gmel.*
The Dartford Warbler, *Lath.*

THIS name is, in Provence, applied to a very ſmall bird, which appears to us more related to the fauvettes than to any other genus. Its total length is five inches, of which the tail takes up near the one half. It probably received this appellation becauſe it conceals itſelf among cabbage *(chou)*: it ſearches for the young butterflies that are bred on the leaves, and in the evening it ſquats and hides itſelf from its enemy, the bat, which roves above its cold lodging. But ſeveral perſons have aſſured me, that *pitchou* has no relation to *chou*, and ſignifies only *little* or *ſlender*; which agrees with Italian etymology *, and ſuits well this bird, which is almoſt as ſmall as a wren.

The bill of the pitchou is long in compariſon to its body, being ſeven lines: it is blackiſh at the tip, whitiſh at the baſe: the upper mandible is ſcalloped near the end: the wing is very

* *Piccino*, *Piccinino*.

ſhort, and covers only the origin of the tail: the *tarſus* is eight lines: the nails are very thin, and the hind one is the largeſt: all the upper part of the body, from the forehead to the end of the tail, is deep cinereous: the quills of the tail, and the great quills of the wings, are edged with light cinereous on the outſide, and blackiſh within.—We are indebted to M. Guys of Marſeilles for our knowledge of this bird.

FOREIGN BIRDS

RELATED TO THE FAUVETTES.

I.

THE SPOTTED FAUVETTE,

FROM THE CAPE OF GOOD HOPE.

Motacilla Africana, *Gmel.*
The African Warbler, *Lath.*

THIS bird, deſcribed by Briſſon, is one of the largeſt, ſince he makes it equal to the brambling, and ſeven inches three lines long. The crown of the head is rufous, variegated with blackiſh ſpots in the middle of the feathers: the top of the neck, the back and the ſhoulders are clouded, except that their edge is dirty gray: near the rump, on the coverts of the wings, and the upper ſurface of the tail, they are edged with rufous: all the under and fore part of the body is ruſty white, variegated with ſome blackiſh ſpots on the flanks: on each ſide of the throat there is a ſmall black ſtripe: the quills of the wings are brown, with the outer border rufous: the four quills in the middle of the tail are ſimilar, the reſt are rufous, but all

 of

of them are ſharp and pointed: the bill is horn-colour, and eight lines long: the feet are ten lines, and of a dun gray.

II.

THE SMALL SPOTTED FAUVETTE,

FROM THE CAPE OF GOOD HOPE.

THIS is a new ſpecies, and introduced by Sonnerat: it is ſmaller than the babbler fauvette, and its tail is longer than its body: the whole of its robe is brown, and the breaſt is ſpotted with blackiſh on a yellowiſh white ground.

III.

THE SPOTTED FAUVETTE,

FROM LOUISIANA.

Motacilla Noveboracenſis, *Gmel.*
The New York Warbler, *Penn.* & *Lath.*

IT is of the ſize of the tit-lark, and reſembles it in the manner in which all the under part

part of the body is ſpotted with blackiſh on a yellowiſh white ground ; theſe ſpots reach from near the eyes to the ſides of the tail : a ſtreak of white riſes at the angle of the bill, and terminates in the eye : all the upper ſurface, from the crown of the head to the end of the tail, is mixed with cinereous and deep brown.

We ſhould not have heſitated to refer to this ſpecies, as a variety proceeding from age or ſex, another fauvette which was alſo ſent from Louiſiana, of which the plumage is a lighter gray, and has only a few traces of the ſpots which are diſtinctly painted on the former : the upper part of the body is whitiſh ; a veſtige of a yellowiſh tinge appears on the ſides, and the rump : beſides, theſe two birds are of the ſame ſize ; the quills and the great coverts of the wings in the laſt are fringed with whitiſh ; but an eſſential difference takes place in their bills : in the firſt, it is as large as the reed fauvette, and in the ſecond, it is hardly equal to that of the ſmall fauvette. This diverſity in the principal part appears to be ſpecific, and we ſhall therefore conſtitute this another ſpecies, under the name of SHADED FAUVETTE FROM LOUISIANA*.

* Motacilla Umbria, *Gmel.*
The Umbroſe Warbler, *Penn.*
The Duſky Warbler, *Lath.*

IV. THE

IV.

THE YELLOW-BREASTED FAUVETTE,

FROM LOUISIANA.

THIS is one of the handſomeſt and moſt brilliant of the whole genus : a half-maſk of black covers the face and temples even beyond the eyes, and ſupports a white border ; all the upper ſurface is olive, all the under part yellow, with an orange tint on the ſides. It is of the ſize of the paſſerine warbler. It was brought from Louiſiana by Lebeau.

A fourth ſpecies is the *Greeniſh Fauvette* from the ſame country. It is of the ſize of the ſpotted fauvette, which we have juſt deſcribed : its bill is as long, and is ſtronger : its throat is white ; the under part of the body white gray ; a white ſtreak paſſes below the eye, and beyond it : the crown of the head is blackiſh ; the upper ſide of the neck is deep aſh colour ; the flanks and the back are greeniſh, on a light brown ground ; a purer greeniſh borders the quills of the tail, and the outſide of thoſe of the wings, whoſe ground is blackiſh. It ſeems, by reaſon of its blackiſh hood, to form the correlative to our black cap, which it equals in ſize.

V. THE

V.

THE RUFOUS-TAILED FAUVETTE,

FROM CAYENNE.

Motacilla Ruficauda, *Gmel.*
The Rufous-tailed Warbler, *Lath.*

ITS total length is five inches one fourth: it has a white throat, encircled with rufty dotted with brown: the breaft is light brown: the reft of the under part of the body is white, with a rufty tinge on the inferior coverts of the tail: all the upper fide, from the crown of the head to the origin of the tail, is brown, with a rufous tinge on the back; the coverts of the wings are rufous, their quills edged exteriorly with rufous; and all the tail is of that colour.

VI.

THE FAUVETTE OF CAYENNE,

WITH A BROWN THROAT AND YELLOW BELLY.

Motacilla Fufcicollis, *Gmel.*
The Yellow-bellied Warbler, *Lath.*

THE throat, the upper fide of the head, and of the body, are of a greenifh brown: the quills and

and coverts of the wings have the ſame ground colour, but are edged with ruſty; thoſe of the tail with greeniſh: the breaſt and belly are yellow, ſhaded with fulvous. It is one of the ſmalleſt of the genus, and ſcarcely exceeds the willow-wren: its bill is broad, and flat at its baſe, and in that reſpect it appears to reſemble the fly-catchers, which are in fact nearly related to the fauvettes, being diſtinguiſhed only by ſlight differences of conformation, while they are connected by one leading character, viz. that their modes of living are the ſame.

VII.

THE BLUEISH FAUVETTE OF SAINT DOMINGO.

Motacilla Cærulefcens, *Gmel.*
The Blue-grey Warbler, *Lath.*

THIS pretty little fauvette is only four inches and a half long; and all the upper ſide of the head, and of the whole of the body, is blue cinereous: the quills of the tail are edged with the ſame colour, on a brown ground: there is a white ſpot on the wing, of which the quills are brown: the tail is black: the reſt of the under ſide of the body is white.

We are ſorry that we know nothing of the habits of theſe different birds. Nature ſtamps every animated being with inſtincts and powers ſuited to their climates, and as various as thoſe: ſuch ſubjects are always worthy of being obſerved, but almoſt always want proper obſervers. Few are ſo intelligent or ſo laborious as the perſon * to whom we owe the intereſting account of another little fauvette in St. Domingo, called the *yellow-neck* in that iſland.

* M. le Chevalier Lefevre Deſhayes.

THE

YELLOW-NECK.

Motacilla Penſilis, *Gmel.*
Sylvia Penſilis, *Lath. Ind.*
The Penſile Warbler, *Lath. Syn.*

SUCH is the name *(cou-jaune)* which the ſettlers in St. Domingo have beſtowed on a ſmall bird*, which to beauty of plumage joins an eaſy ſhape and a pleaſant warble: it ſits upon the trees which are in bloſſom, and ſtrains its little throat: its voice is ſlender and weak, but varied and delicate; each paſſage of its muſic is compoſed of rich and full cadences†. The bird is the more charming, as its ſong laſts not only ſpring, the ſeaſon of love, but is prolonged through almoſt all the months of the year. We ſhould almoſt ſuppoſe that its paſſion ſuffers no

* They alſo call it *the goldfinch*: yet the yellow-neck has the ſlender bill of the pettychaps, or red-breaſt, and the port, the temper, and habits of the latter; nor has it any thing analogous to the goldfinch but the warble, which is alſo very different.

† "The ſong of the *corn or cane bird* reſembles in the thinneſs of its tones, and the quality of its modulation, the warble of the yellow-neck." *Note of M. Lefevre Deſhayes*, an ingenious and ſenſible obſerver, to whom we owe the details in this article, and many other intereſting facts in the natural hiſtory of the birds of St. Domingo.

intermiſſion; and, in that caſe, it might be inceſſantly fired to warble its amorous tale. As ſoon as the weather grows fine, eſpecially after thoſe ſudden and exceſſive torrents of rain which are ſo frequent in the Weſt Indies, the male tunes his voice, and chants whole hours together: the female alſo ſings; but her notes are neither ſo well ſupported, nor ſo finely blended.

Nature, who paints moſt of the birds in the New World with the richeſt colours, denies them the charms of ſong, and, in the deſert tracts, ſhe beſtows only ſome ſavage cries. The yellow-neck is one of the ſmall number whoſe warble is lively and cheerful, and whoſe plumage is at the ſame time diſtinguiſhed for beauty: the tints are well blended, and are heightened by the fine yellow which ſpreads over the throat, the neck, and the breaſt: black gray predominates on the head, and, growing more dilute as it deſcends to the neck, it changes into a deep gray on the back: there is a white line which crowns the eye, and joins to a ſmall yellow ſtreak lying between the eye and the bill: the belly is white, and the ſides are ſpeckled with white and black gray: the coverts of the wings are ſpotted with black and white, diſpoſed in horizontal ſtripes; there are alſo large white ſpots on the quills, of which there are ſixteen in each wing, and with a ſmall white gray border at the

the end of the great webs: the tail consists of twelve quills, of which the four outer ones are marked with large white spots: a scaly fine skin, of a greenish gray, covers the legs: the bird is four inches and nine lines in length; its alar extent eight inches, and it weighs one gros and a half.

Under this rich clothing, the pensile warbler has the figure and proportions of the fauvettes; and its habits are also the same. It prefers for its haunts the sides of rivulets, and the cool refreshing spots near springs, and wet gullies; whether because a mild temperature is most congenial to its nature, or that it seeks retirement where nothing may disturb its music. It flutters from tree to tree, and from branch to branch, and warbles in its passage through the air. It preys on flies, caterpillars, and butterflies; and yet, in the season, it cracks the seeds of the guava and water melon, &c. probably to find the maggots which are bred in these at a certain state of maturity. It appears neither to arrive in St. Domingo nor depart: its flight, though rapid, is not so lofty, nor so continued, as to waft it over the ocean *, and it may be regarded as a native of that island.

But

* M. Deshayes compares the flight of the yellow-neck to that of the bird called at St. Domingo *de la Toussaint* (*All-*

But the beauty and ſenſibility of this bird are no leſs remarkable than the ſagacity it diſplays in building and placing its neſt. It does not fix it at the forking of the branches, as uſual with moſt other birds; it ſuſpends it to binders hanging from the netting, which they form from tree to tree, eſpecially thoſe which fall from branches leaning over the rivers and deep ravines: the neſt conſiſts of dry blades of graſs, the ribs of leaves, and exceedingly ſmall roots, interwoven with the greateſt art; it is faſtened, or rather it is worked into the pendent ſtrings; it is really a ſmall bed rolled into a ball, ſo thick and compacted as to exclude the rain, and which rocks in the wind without receiving any harm.

But the elements are not the only enemies againſt which this bird has to ſtruggle: with wonderful ſagacity it provides for its protection from other foes: the opening is not made on the top or ſide of the neſt, but at the bottom; nor is the entrance direct: after the bird has made its way into the veſtibule, it muſt paſs another aperture before it deſcends into the abode of its family: this lodgment is round and ſoft, and lined with a ſort of lichen which grows on the trees, or

(*All-Saints*), ſeemingly becauſe it arrives about that time. "It is nearly of the ſize," ſays he, "of the yellow-neck; but this is very delicate in compariſon, and the muſcles of its wings are much leſs vigorous than in the bird *de la Touſſaint*."

with the ſilk of a plant called by the Spaniards *mort à cabaye* *.

By this laborious conſtruction, the young brood are protected againſt the attacks of the rapacious birds, and of the rats and ſnakes. Yet dangers ſtill await them: when they are about to fly, many are devoured by the owls and rats, and the ſpecies ever remains limited. Such is the fate of the weak and gentle creatures in thoſe regions, where the noxious kinds ſpread and prevail by their numbers.

The female lays only three or four eggs; ſhe hatches more than once in the year, but how often is not known: the young ones are ſeen in the month of June, and ſome are ſaid to appear as early as March, and others are found in the end of Auguſt, or in September; they ſoon leave their mother, but never rove far from the place of their nativity. [A]

[A] Specific character of the *Motacilla Penſilis*: "It is gray, below yellow, its belly and eyebrows white, its ſtraps ſpotted with yellow, the coverts of its wings marked with alternate ſtripes of black and white."

* "It is a plant which grows in the ſavannas of St. Domingo, and delights in humid ſituations: its milk is a ſtrong poiſon, which is no doubt the reaſon of its name, *mort à cabaye*." *Note de M. le Chev. Deſhayes.*

THE

THE

REDSTART.

Le Rossignol de Muraille, *Buff.**
Motacilla-Phœnicurus, *Linn. & Gmel.*
Phœnicurus, *Briss. Frisch, &c.*
Ruticilla, *Ray. Will. Briss. Klein, Gibb. &c.*
Pyrrhulas, *Johnst.*

THE song of this bird has neither the extent nor the variety of the nightingale's warble; but it partakes of the same modulations, and wears an air of tenderness and melancholy. Such at least are the emotions which this awakens in us; for, with regard to the bird itself, it must be the expression of joy and pleasure, as it is the expression of love, which is equally delicious to every animated being. This is the only analogy that subsists between the two

* In Greek, Φοινικȣρος, Arist. *Hist. Anim.* lib. ix. 49: in Latin, *Phœnicurus*, Plin. lib. x. 29: in Italian, *Codirosso, Corossolo, Revezol:* In German likewise its names denote the reddish colour of its tail; *Rot-stertz, Rot-schwentzel, Wein-vogel, Rot-schwantz, Schwantz-kehlein*, and the female *Roth-schwentzlein.* It is also called *Hauss-roetele, Summer-roetele* (house or summer red bird): in Silesian, *Wustling:* in Prussian, *Saulocker:* in Polish, *Czerwony Ogonek.* The English name Redstart is evidently borrowed from the German *Rot-startz*, which signifies *red-tail.*

birds; their habits, their ſize, their plumage* are different, though in French the ſame generic name of *nightingale* has been uſually applied to both.

This bird appears with the reſt in the ſpring, and ſits on towers and the ruins of deſerted buildings, and there it pours forth its notes. It even procures ſolitude in the midſt of cities, where it ſettles on the top of a high wall, in a belfry, on a chimney, &c. always ſeeking the moſt lofty and moſt inacceſſible ſpots: it is alſo found in the heart of the thickeſt foreſts. It flies nimbly; and when it perches it vents a feeble cry †, and quivers its tail inceſſantly, not upwards and downwards, but horizontally, from right to left. It prefers the mountainous tracts, and ſeldom viſits the plains ‡. It is much ſmaller than the nightingale, and even ſomething ſmaller than the redbreaſt; its form is more ſlender, and longer; a black horſe-ſhoe covers its throat, the fore part and ſides of its neck; the ſame black encircles its eyes, and reaches under its bill; a white bar maſks its face: the crown and back of its head, the upper part of its neck and back, are of a gloſſy, but deep gray: in ſome ſubjects, probably old ones, this gray is almoſt black: the wing quills are blackiſh cinereous; their outer webs are of

* Belon. † Id. ‡ Olina.

a lighter

a lighter caſt, and fringed with whitiſh gray: below the black horſe-ſhoe, a fine rufous fire colour decorates a great part of the breaſt; and, fading ſomewhat on the ſides, it again reſumes its luſtre on all the plumage of the tail, except the two middle feathers, which are brown; the belly is white, and the feet black; the tongue is forked at the end, as in the nightingale *.

The female differs ſo much from the male, that ſome authors have reckoned it a ſecond ſpecies †: it has neither the white face nor the black throat of the latter; both theſe parts are gray mixed with ruſty, and the reſt of the plumage is of a lighter tinge.

Theſe birds breed both in towns and in the country, in hollow trees or in the crags of rocks: they lay five or ſix blue eggs: the young are hatched in May ‡. During the whole time of incubation, the male chants from ſome neighbouring eminence, or from the top of a detached building §; and his muſic is ſofteſt at day-break ‖.

It is ſaid that theſe birds are timorous and ſuſpicious, and that they will abandon their neſt, if they be ſeen employed in conſtructing it, and that they will deſert the eggs if they be touched. All this is probable; but what Albin adds is

* Belon. † Linnæus and Klein. ‡ Schwenckfeld.
§ Olina, *Uccell.* p. 47. ‖ Aldrovandus, t. ii. p. 750.

absſurd; that if the young be handled, the parents will leave them to their fate, or throw them out of their neſt *.

The redſtart, though it lives amidſt our dwellings, continues ſtill ſavage. It has neither the familiarity of the redbreaſt, the ſprightlineſs of the fauvette, nor the animation of the nightingale; its habits are ſolitary, its character is ſullen and ſad †. If it be caught in the adult ſtate, it will refuſe all ſuſtenance, and pine to death ‡; or if it ſurvive the loſs of its liberty, an obſtinate ſilence will mark its diſconſolate condition. However, if it be taken from the neſt and raiſed in the cage, it will ſing; and inſtructions, or the imitation of other birds, will improve its warble §, which is heard indiſcriminately at every hour, and even during the night ‖.

It is fed with crumbs of bread, and with the ſame paſte as the nightingale; it is even more delicate ¶. When at liberty, it lives on flies,

* Albin, vol. i. p. 44.

† "Their young much reſemble thoſe of the redbreaſts; they cannot be ſo eaſily raiſed. I have kept one a whole winter; it ſeemed of a timid diſpoſition, yet was it continually hopping, and had a very keen eye; it could diſtinguiſh at one end of the room the ſmalleſt inſect at the other, and darted to it in an inſtant, emitting a cry in ſeizing it." *Note communicated by the Viſcount de Querhoënt.*

‡ Albin, vol. 1. p. 44. § Idem, ibidem.

‖ Olina, *Uccelleria*, p. 47. ¶ Belon.

ſpiders,

ſpiders, the chryſalids of ants, ſmall berries, and ſoft fruits. In Italy it pecks the figs; and Olina tells us that it is ſeen in that country as late as the month of November, though in France it diſappears in October. It departs when the redbreaſt begins to viſit our habitations; and this is the reaſon perhaps that Ariſtotle and Pliny aſſert that the redbreaſt of winter, and the redſtart of ſummer, are the ſame bird *. Even in their mi rations, the redſtarts ſhew their ſolitary diſpoſition; they never aſſemble in flocks, but arrive and depart ſingly †.

There are varieties of the redſtart; ſome derived from climate, others occaſioned by age. Aldrovandus mentions three; but the firſt is a female, and the ſecond is an imperfect figure from Geſner, and only the bird diſguiſed; the third only is a true variety; it has a long white ſtreak on the fore part of the head: this is what

* Ariſt. *Hiſt. Anim.* lib. ix. 49.—Plin. lib. x. 29—Belon, *Nat. des Oiſeaux*, p. 347, 348.

† "This year I took a walk into the park one day when there was probably a numerous flight, for I ſprung them every minute from the hedge-rows, and almoſt always one by one. I got ſo near many of them as to diſtinguiſh them eaſily: it was about the 15th of September. Theſe birds are very common at Nantua in the ſpring and ſummer, and probably leave the mountains in the beginning of autumn, but without ſettling in our plains, where it is very rare to ſee them at any other ſeaſon." *Note communicated by M. Hebert.*

Briſſon calls the *cinereous redſtart*, and which Willughby and Ray deſcribe from Aldrovandus. Friſch mentions another variety of the hen redſtart, in which the breaſt is marked with rufous ſpots; and this variety conſtitutes Klein's ſecond ſpecies. The gray redſtart of Edwards *, ſent from Gibraltar to Cateſby, and which Briſſon makes his ſecond ſpecies, is probably only a variety of climate. It is of the ſame ſize with the common redſtart; the greateſt difference is, that there are no rufous tints on its breaſt, and that the outer edges of the middle quills of its wing are white.

Another variety nearly the ſame is the bird ſent to us by M. D'Orcy, in which the black colour of the throat ſpreads over the breaſt and ſides; whereas in the common redſtart theſe parts are rufous. We do not know whence M. D'Orcy received it: it had a white ſpot on the wing, of which the quills are blackiſh; all the cinereous caſt of the upper part of the body is deeper than in the redſtart, and the white of the forehead is much leſs apparent.

There is beſides in America a ſpecies of redſtart deſcribed by Cateſby, which we ſhall leave undecided, and not ranged expreſsly with that of Europe; not ſo much becauſe of the difference of characters, as of the wide ſeparation be-

* Motacilla Gibraltarienſis, *Gmel.*

tween

tween the continents. In fact, Catesby ascribes to the Virginian redstart the same habits which we survey in our own. It lives in the closest woods; it is seen only in summer: its head, neck, back and wings are black, except a small spot of vivid rufous on its wing; the rufous colour of the breast is divided into two by the continuation of the gray of the stomach; the point of its tail is black. Are these differences specific, and more marked than what might be expected from the influence of another hemisphere?

The *Bugey-collier (charbonnier du Bugey)*, according to Hebert's account, is also the redstart *. We shall make the same assertion in regard to the *russet-tail* † of Provence, of which

* "I think that the name of redstart *(queue-rouge)* may also be given to a bird of the bulk of a pettychaps, which is very common in Bugey, and there called the *collier (charbonnier)*: it appears both in the towns and among the rocks; it nestles in the holes. Every year it has a nest on the ridge of the house which I occupy, in a hole at a great height: while the hen covered, the cock perched very near her on some point of the ridge, or on some very lofty tree, and repeated incessantly a doleful warble, which had only two variations, succeeding constantly in the same order at equal intervals. These birds have a sort of convulsive trembling of the tail. I have seen them sometimes at Paris in the Tuilleries, never in Brie, nor have I heard their warble in Bugey." *Note communicated by M. Hebert, Farmer General at Dijon.*

† *Cul-rousset*, ou *Cul-rousset farnou.*

we

we have been informed by M. Guys. We likewise ſuppoſe that the * *chimney-bird* of the ſame province is the redſtart; at leaſt, the analogy of habits and alſo reſemblance of characters ſeem to evince the identity. [A]

[A] Specific character of the Redſtart, *Motacilla-Phœnicurus*: "Its throat is black, its belly and tail rufous, its head and back hoary." The redſtart is frequent in England: its ſong is ſomewhat like that of the white-throat, though ſuperior. Its egg is blue.

* *Fourmeirou*, ou *fourneirou de cheminee*.

THE

RED-TAIL.

Rouge-Queue, *Buff.*
Motacilla-Erithacus *, *Linn. & Gmel.*
Sylvia-Erithacus, *Lath. Ind.*
Phœnicurus Torquatus, *Briſſ.*
Phœnicurus Alter, *Aldrov.*
The Gray Redſtart, *Penn.*

ARISTOTLE mentions three ſmall birds, and marks by the compoſition of the names which he applies that the principal feature of their plumage is a flame tint. Theſe are the Φοινικȣρος, which Gaza tranſlates by *ruticilla*; Ερυθακος, tranſlated *rubecula*; and Πυρρȣλας, which he renders *rubicilla* †. We are pretty confident that the firſt is the redſtart, and the ſecond the redbreaſt; indeed the habits which Ariſtotle aſcribes to theſe, that the former lives in the ſummer near our habitations, and diſappears in autumn when the latter arrives, can belong to no other birds which have

* *Erithacus* might properly perhaps be written *Erythacus*. T.

† Φοινικȣρος is derived probably from Φοινιξ the Tyrian purple, and ȣρα a tail; Πυρρȣλας is evidently formed from πυρ fire; Ερυθακος from ερυθος red.

a rutilous

a rutilous plumage, but the redſtart or red-breaſt. It will be more difficult to aſcertain the Πυρρȣλας or *rubicilla*. Theſe names have been applied by all the nomenclators to the bulfinch: their opinion was noticed, but not diſcuſſed, at that article; and we ſhall now reſume the ſubject, and ſtate the reaſons which diſpoſe us to make a very different concluſion.

Ariſtotle enumerates at this place the ſmall birds, with a ſlender bill, which live chiefly on vegetables; ſuch are, ſays he, the *cygalis*, the *beccaſico* (or epicurean warbler), the *melancoryphus** (or black-cap), the *pyrrhulas*, the *erythacus*, the *hypolaïs* (the babbling warbler): but I aſk whether the bulfinch can be claſſed with theſe; or is not that bird the moſt decidedly granivorous? It will not touch inſects in the ſeaſon when moſt others feed upon them; and

* I know that Belon and many naturaliſts after him, have referred alſo to the bulſinch the name of *melancoryphus*; but I am convinced that this application is erroneous. Ariſtotle ſpeaks in two places of the *melancoryphus*, and in both he alludes to two different birds, neither of them the bulfinch: in the firſt paſſage we ſhall prove that he means the *pyrrhulas*; in the ſecond, it is ſaid *to lay twenty eggs, to neſtle in hollow trees, and to feed on inſects*, which character is true only of the black-headed titmouſe.—This little diſcuſſion ſeemed to me the more neceſſary, as Belon has of all the naturaliſts diſcovered the moſt ſagacity in referring the ancient names, and as the nomenclature is exceedingly embarraſſing.

it

it ſeems to differ as much from the vermivorous birds by its inſtincts as by the ſhape of its bill: and it is not likely that Ariſtotle would overlook this circumſtance.

To what other bird, then, can we aſcribe theſe properties? I perceive none but the *red-tail*, which inhabits the woods with the red-breaſt, and alſo feeds on inſects during the whole ſummer, and departs at the ſame time in winter. Wotton conceived that the pyrrhulas was a kind of red-tail *, and Johnſton makes the ſame remark †; but the former was miſtaken in ſuppoſing this bird to be the redſtart, ſince Ariſtotle nicely diſtinguiſhes them.

The red-tail is actually very different from the redſtart, and Ariſtotle and Geſner did well to ſeparate them. It is larger than the redſtart: it never viſits our dwellings, nor neſtles in the walls; but lives in the woods and buſhes like the fauvettes and beccafigos: its tail is of a light vivid fire colour; the reſt of the plumage conſiſts of gray, eſpecially on all the upper ſurface, and deeper and fringed with ruſty on the quills of the wings, and with white gray, mixed confuſedly with ruſty, on all the fore part of the body: the rump is rufous, and ſo is the tail. Some have a fine black collar, and

* *Apud Geſnerum*, p. 701.
† Pyrrhulas, *Johnſt.*

the

the colours over the whole of their plumage more lively and varied. Briſſon has formed theſe into a ſecond ſpecies *; but we conceive that they are only the males, and ſome very experienced bird-fanciers have confirmed our opinion. Briſſon ſays, that the collared red-tail is *found in Germany*, as if it were peculiar to that country: but wherever the gray red-tails occur, the collared red-tails are equally frequent. Beſides, he is miſtaken in his reference; for the figure in Friſch, which he ſuppoſes to be the red-tail, is only the female of the blue-throated warbler.

We regard the collared red-tail, therefore, as the male, and the gray red-tail as the female: in both, the tail is equally red; but, beſides the collar, the male has a deeper plumage, being a brown gray on the back, and gray ſpotted with brown on the breaſt and ſides.

Theſe birds prefer the hilly countries, and ſcarcely appear in the low grounds, except in their autumnal paſſage †. They arrive in the month of May in Burgundy and Lorraine,

* *Phœnicurus Torquatus.*

† "I have ſeen in Brie, in autumn, a bird which likewiſe had a very rufous tail, but different from the red-ſtart: I thought it was the ſame with the *collier* of Nantua in its firſt year. Almoſt all theſe birds change their colour the firſt moulting, and all the birds which feed on inſects are ſubject to migrations in autumn." *Note communicated by M. Hebert.*

and

and ſoon bury themſelves in the woods, where they remain all the ſummer. They breed in the ſmall buſhes near the ground; their neſt conſiſts of moſs, and lined with wool and feathers; it is of a ſpherical ſhape, with its opening facing the eaſt, and the moſt ſheltered from the ſtormy winds; it contains five or ſix eggs, variegated with gray.

The red-tails leave the woods in the morning, and return to avoid the heat of the day; they emerge in the evening, and appear in the neighbouring fields, but again return to their abodes to paſs the night. Theſe habits, and many other features of reſemblance, ſeem to imply that they belong to the genus of the redſtart. The red-tails, however, have neither its ſong nor its warble; they utter only a feeble mellow note *sŭit*, and it dwells on the firſt ſyllable, and ſpins it out with great ſoftneſs. They are in general ſilent, and exceedingly compoſed*: if a ſmall detached branch projects from a buſh or ſtretches acroſs a path, they will alight on it,

* A red-tail taken in autumn and let looſe in an apartment, made not the ſmalleſt cry, whether flying, walking, or reſting. Shut in the ſame cage with a pettychaps, the latter darted every minute againſt the bars, while the former remained ſtill whole hours in the ſame place, where the pettychaps dropped upon it at each ſpring; and it ſuffered itſelf to be thus trod upon all the time the pettychaps lived, that is, about thirty-ſix hours.

2 making

making a ſlight vibration with their tail like the redſtart.

It anſwers the decoy, but does not ruſh on with the vivacity and eagerneſs of other birds; it ſeems only to follow the crowd. It is alſo caught near ſprings about the end of autumn; and it is then very fat, and has a delicate flavour. Its flight is ſhort, and reaches only from buſh to buſh.—Theſe birds depart in the month of October; they may for ſeveral days be ſeen following each other along the hedges; but after that time none of them remain in France. [A]

[A] Specific character of the Red-tail, *Motacilla-Erithacus*: "Its back and its wing quills are cinereous, its belly and tail quills rufous, the two outermoſt cinereous." It is unknown in Britain.

THE GUIANA RED-TAIL.

Motacilla Guianenſis, *Gmel.*

THE quills of the wings are of the ſame rufous colour as thoſe of the tail; the back is gray, and the belly is white. We are unacquainted with its habits and economy; but it is probably related to the European kind, and its inſtincts nearly the ſame.—We received it from Cayenne.

THE

EPICUREAN WARBLER.

Le Bec-Figue, *Buff.*
Motacilla-Ficedula, *Linn.* & *Gmel.*
Sylvia-Ficedùla, *Lath. Ind.*
Ficédula, *Aldrov. Gesner, Johnst.* &c.
Beccafiço, *Olina,* & *Russel. Aleppo* *.

THIS bird, which, like the ortolan, is esteemed by the luxurious as the highest delicacy, is not remarkable for its beauty. All its plumage is of a dull colour; the only shades are gray, brown, and whitish, to which the blackish cast of the wing quills is added, without heightening the complexion: a white spot, which transversely intersects the wing, is the most conspicuous touch of its colours, and that which many naturalists have assumed for the character †: the back is brown gray, which begins on the crown of the head, and spreads to the rump; the throat is whitish; the breast lightly tinged with brown: the belly is white, and so are the outer webs of the

* In Greek Συκαλις, from συκον a fig: the Latin name, *Ficedula,* is formed from *ficus* a fig, and *edo* to eat: in German, *Grasz-mach* or *Wustling.*

† Frisch.—Brisson.—Aldrovandus.

two firſt quills of the tail; the bill is ſix lines in length, and ſlender. The alar extent is ſeven inches, and the total length of the bird is five: in the female, the colours are all duller and paler than in the male *.

Theſe birds are natives of the ſouth, and ſeem to viſit our climate only to wait the maturity of the figs; they arrive late in the ſpring, and diſappear before the earlieſt froſts of autumn. However, they ſpread in the ſummer over a great extent in the northern countries; for they are found in England †, in Germany ‡, in Poland §, and even in Sweden ‖. They return in autumn to Italy and Greece, and probably advance to winter in ſtill warmer regions. They ſeem to change their habits with their climate; for, in the countries of the ſouth, they appear in flocks; while, in the temperate climates, they remain always diſperſed: they inhabit the woods, feed on inſects, and live in ſolitude, or rather in the endearing ſociety of their female. Their neſts are ſo artfully concealed, that it is very difficult to diſcover them ¶: during that ſeaſon, the male

* Aldrovandus. † Willughby. ‡ Klein.
§ Rzaczynſki. ‖ Linnæus.

¶ "The beccafigo neſtles in our foreſts, and, to judge from analogy, in the holes of trees at a great height above the ground, like the collared fly-catchers: for this reaſon, they male ſits on the ſummit of ſome tall tree, and makes a feeble chirping, which is unpleaſant, and much like that of the wheat-ear. The epicurean warblers arrive in Lorraine in April, and diſappear in Auguſt, and ſometimes later *. In that province, they are called *mulberry-birds*, or *little wood-finches* †, which has tended to miſlead; for, at the ſame time, the name of *fig-pecker* ‡ has been beſtowed on the tit-lark, which is of a very different ſpecies. Nor are theſe the only miſtakes that have been made in regard to the name. Belon preſumes, becauſe the bulfinch in Italy ſeems fond of figs, that it is the ſame bird with what is called in that country *beccafico*, and he infers that it is the real *ficedula*, to which Martial alludes. But the bulfinch is as different from the epicurean warbler in the taſte of its fleſh, which is bitter, as in the ſhape of its bill, in its colours, and in the reſt of its figure. In our ſouthern provinces, and in Italy, all the different ſpecies of fau-

are very difficult to be found. In 1767 or 1768, having ſeen or heard one of theſe birds ſing, which was perched upon the ſummit of a very tall tree, I watched it attentively, and returned ſeveral times without being able to diſcover the neſt, though I always ſaw the bird again. It had a feeble chirrup like the wheat-ear, by no means agreeable. It perched extremely high, and ſeldom came near the ground." *Note communicated by M. Lottinger.*

* Lottinger. † *Mûriers*, or *petits pinçons des bois.*

‡ *Bec-figue.*

 vettes

vettes are confounded under the name of fig-pecker, and almoſt all theſe birds have a ſlender long bill *; but the true fig-pecker, or epicurean warbler, is alſo well known and diſtinguiſhed for the delicacy of its flavour.

Martial, who aſks why this ſmall bird, which feeds alike on grapes and on figs, did not aſſume the name of the former rather than that of the latter †, would have adopted the appellation which it receives in Burgundy, where we call it *vinette*; becauſe it frequents vineyards, and lives on the grapes. Along with theſe, however, it eats inſects alſo, and the ſeeds of the herb mercury. Its feeble chirp may be expreſſed by *bzi*, *bzi*; it flies by jerks; it walks, but never hops, and it runs along the ground in the vineyards, and mounts upon vines and hedge encloſures.

Though the epicurean warblers never journey till about the month of Auguſt, and never appear in flocks but then, in moſt of our provinces; yet they are ſeen in the middle of ſummer at Brie, where ſome of them probably build their neſts ‡. In their paſſage, they fly in ſmall bodies of five or ſix. They are caught by

* Salerne.

† "Cum me ficus alat, cum paſcar dulcibus uvis,
Cur potius nomen non dedit uva mihi?"
MART.

‡ Hebert.

by the noose, the springe, or the mirror, in Burgundy and along the Rhone, where they make their appearance about the end of August, and in September.

It is in Provence where they are properly named fig-peckers, for they are constantly seen on the fig-trees pecking the ripest fruit. They never leave these, except for the shade and cover of the bushes and of the leafy horn-beam. They are caught in great numbers during the month of September in Provence, and in many islands of the Mediterranean, especially in Malta, where they appear at times in prodigious flocks; and it has been remarked that they are much more numerous during their passage in autumn, than at their return in spring*. The same is the case at Cyprus, where formerly they were articles of traffic: they were sent to Venice, in pots filled with vinegar and odoriferous herbs †. When Cyprus belonged to the Venetians, a thousand or twelve hundred of these pots were furnished

* Chevalier de Mazy.

† *Voyage de Pietro della Valle*, t. viii. p. 153. He adds, that in some places, as at *Agia Nappa*, those who eat beccafigos are sometimes attacked by disorders, owing to the scammony which these birds find in the neighbourhood; they feed also, in the islands of the Archipelago, on the fruits of the lentisc.

 every

every year *; and the fig-pecker, or epicurean warbler, was generally known through Italy by the name of *Cyprus bird (Cyprias, uccello di Cypro)*, which appellation, according to Willughby, was usual even in England.

This delicious bird has long been famous: Apicius mentions it more than once, with the little thrush, as both equally exquisite. Eustathius and Athenæus speak of the fowling for epicurean warblers †, and Hesychius preserves the name of a net with which they were caught in Greece. Nothing indeed is more delicate or juicy than the meat of these birds at the proper season; it is a pellet of light, savoury, melting fat, of easy digestion; it is the extract of all the juices of the excellent fruits on which they feed.

We are acquainted with only one species, though the name has been bestowed on several ‡. If indeed we call every bird *fig-pecker* which pecks

* Dapper's Description of the Archipelago Islands.

† In Gesner.

‡ Aldrovandus gives (t. ii. p. 759) two figures of the fig-pecker, of which the second, according to himself, exhibits only a variety of the first, perhaps even accidental, and which, he says, may be called *the variegated fig-pecker, black and white being mixed through its whole plumage, as the figure shews:* but this figure shews only in that the white on the wing is a little broader, and there is some white on the fore side of the neck and on the breast; which constitutes only an individual variety.

the

the figs in that ſeaſon, the fauvettes and all the birds with ſlender bills, and even ſome with thick bills, would be entitled to that appellation. Hence the Italian proverb, *Nel' meſe d'Agoſto ogni uccello e beccafico* (In the month of Auguſt every bird is a fig-pecker). But this vulgar ſaying, which very well expreſſes the delicacy of flavour that the figs give to the little birds which feed on them, would be a very vague and improper foundation for arrangement; and were it adopted, it would introduce the greateſt confuſion. Yet ſome naturaliſts have fallen into this error. *The hemp fig-pecker* of Olina (*beccafico canapino*) is only the babbling warbler. The great warbler, or the pettychaps, is, according to Ray, called *beccafigo* in Italy. Belon applies the ſame name to the rufous warbler; and we have juſt ſeen that he is ſtill more miſtaken in ſuppoſing that term belonged alſo to the *bulfinch*, to which he is, by conſequence, led to aſſign the ancient appellations of the fig-pecker *ſycalis* and *ficedula*. In Provence, ſeveral different birds are confounded under the name of fig-pecker. M. Guys has, among others, ſent two, which we ſhall ſubjoin to this article, only to ſhew more diſtinctly that they are different birds. [A]

[A] Specific character of the Epicurean Warbler, *Motacilla Ficedula*: "It is ſomewhat brown, below white, its breaſt ſpotted with cinereous." It never appears in England.

THE FIST OF PROVENCE.

THE *fiſt*, ſo called by reaſon of its cry, was ſent from Provence as a ſpecies of fig-pecker, or epicurean warbler. It is entirely different, and reſembles much more the ſky-lark, both in regard to ſize and plumage: the only material difference is, that his hind nail is not long. It chirps *fiſt*, *fiſt*; it does not ſpring when alarmed by noiſe, but runs to cover beneath a ſtone, till the danger is over, which implies that it commonly lives on the ground; a habit the reverſe of that of the fig-pecker.

THE ORTOLAN PIVOTE.

THIS is another bird of Provence, which, though it is called the fig-pecker, is as different from it as the *fiſt*. It is a faithful companion of the ortolans, and always in their train. It reſembles much more the tit-lark, only it is larger, and its hind nail not long.

Nº 126

THE RED-BREAST.

THE

RED-BREAST.

Le Rouge-Gorge, *Buff.*
Motacilla-Rubecula, *Linn. & Gmel. &c.*
Sylvia-Rubecula, *Lath. Ind.*
Rubecula, *Geſner, Briſſ. Klein, &c.*
Erithacus, *Aldrov.*
The Robin Red-breaſt, or Ruddock, *Will.* *

THIS little bird paſſes the whole ſummer in our woods, and viſits our habitations only at its departure in autumn, and its arrival in ſpring: but this laſt appearance is tranſient; it haſtens to the foreſts to enjoy, under the new-ſpread foliage, its ſolitude and its love. Its neſt is placed near the ground, on the roots of young trees, or on herbs that are able to ſupport it; it is conſtructed with moſs, intermixed with hair and oak leaves, and lined with feathers. Often after this is built, ſays Willughby, the bird ſtrews it with leaves, preſerving only a narrow winding entrance under the heap,

* In Greek, Εριθακος, or more properly Ερυθακος: in modern Latin, *Rubecula*: in Italian, *Pettiroſſo, Pettuſſo, Pechietto*: in Portugueſe, *Pitiroxo*: in Catalonian, *Pita Roity*: in German, *Roth-breuſtlein, Wald-roetele, Rot-kropff, Rot-breuſtle, Winter Roetele, Roth-kehlein*: in Saxon, *Rot-kelchyn, Rott-kaehlichen*: in Poliſh, *Gil*: in Illyrian, *Czier-wenka, Zer Wenka*: in Swediſh, *Rot-Gel*.

heap, and even ſhuts the mouth of it with a leaf when ſhe goes abroad. The neſt generally contains between five and ſeven eggs, of a brown colour. During the whole time of incubation, the cock makes the woods reſound with his light, tender ſong: it is a ſweet, delicate warble, enlivened with ſome brilliant modulations, and broken by graceful melting accents, which ſeem to be the expreſſions of amorous deſire. The ſoft ſociety of the female fills up his wiſhes, and makes him uneaſy at the intruſion of other company. He keenly chaſes all the birds of his own ſpecies, and drives them from his little ſettlement; and never did the ſame buſh lodge two pairs of red-breaſts, as faithful as they are amorous *.

The red-breaſt prefers thick ſhade and watered ſpots; it feeds in the ſpring on worms and inſects, which it hunts ſkilfully and nimbly. It flutters like a butterfly about a leaf where it ſees a fly; and on the ground it advances by ſmall ſprings, and darts on its prey, clapping its wings. In autumn, it alſo eats bramble-berries; and, on its departure, it feeds on the grapes of the vineyards, and on the ſorbs in the woods; which is the reaſon that it is apt to be caught in the gins ſet for the thruſhes, and which are baited with theſe ſmall wild fruits. It reſorts often to the ſprings, to bathe or drink,

* *Unum arbuſtum non alit duos erithacos.*

but

but moſt frequently in the autumn; for it is fatter at that ſeaſon than at any other, and needs more to be cooled.

No bird is earlier awake than the red-breaſt; it begins the muſic of the woods, welcomes the dawn of day. It alſo protracts its warble to the lateſt hour, and is ſeen fluttering about in the evening. It is often caught in gins after there is ſcarce light ſufficient for taking it up. It has little ſhyneſs; and its volatility, its reſtleſſneſs, or its curioſity, hurry it into every ſort of ſnare *. It is always the firſt bird that is caught by the decoy; even the voice of the fowlers, and the noiſe made in cutting the branches, attract it; and it alights behind them, and is entangled by the ſpringe or limed twig, the inſtant they are ſet. It anſwers equally the ſcream of the brown owl, or the ſound of the ſlit leaf of ivy †. Their feeble cry *uîp*, *uîp*, whiſtled on the finger, or the chirping of ſome other bird, is ſufficient to put all the red-breaſts round in motion. They fly to the ſpot, ſound-

* Of all birds that live in the ſtate of liberty, the red-breaſt is perhaps the leaſt ſhy: they often approach ſo near, that a perſon might think that he could catch them with the hand; but the bird keeps conſtantly retiring as he advances. It ſeems fond too of accompanying travellers through the foreſts, and it is often obſerved to precede or follow them a pretty long time. *Note communicated by the Sieur Trecourt.*

† The French bird-catchers call this *froûet.*

ing

ing from a diſtance *tirit*, *tiritit*, *tirititit*, with a ſonorous ringing which is not their modulated air, but what they chirp in the morning and evening, and whenever they are excited by a new object. They bruſh eagerly through the whole of the call-ground, till they are ſtopped by ſome of the limed twigs, which are placed in the avenues, or faſtened to poles that are made low on purpoſe to intercept their flight, which is ſeldom more than four or five feet from the ſurface: if one diſentangles itſelf, it makes a third ſmall cry *ti-i*, *ti-i*, which alarms the reſt, and ſtops their further approach. They may alſo be caught in the open parts of the woods by means of poles, to which are faſtened nooſes and limed twigs; but the ſpringes are the moſt ſure and ſucceſsful; nor is it requiſite that theſe be baited; we need only ſet them in the edge of glades, or in the middle of paths, and the unfortunate little bird, puſhed on by curioſity, will throw itſelf into the ſnare.

Wherever large foreſts ſpread, the red-breaſts are found in abundance. In Burgundy and Lorraine particularly theſe birds, which are excellent food, are obtained in the greateſt numbers: many are alſo caught in the neighbourhood of the ſmall towns of Bourmont, Mirecourt, and Neufchâteau; and they are ſent from Nancy to Paris. That province which is well wooded and watered, maintains a vaſt a vaſt variety of birds; its ſituation too, bounded by the Ardenne on the one ſide, and ſkirted by the foreſts of Suntgau, which join the Jura, on the other, is exactly in the direction of the migrations; and for this reaſon, the birds are moſt frequent in the time of their paſſage*. The red-breaſts in particular are brought in immenſe quantities from the Ardennes, where Belon ſaw numbers caught in the ſeaſon.—This ſpecies is diffuſed through the whole extent of Europe, from Spain and Italy to Poland and Sweden; and in every country, theſe little birds prefer the mountains and the woods, to breed and ſpend the ſummer.

The young ones, previous to the firſt moult, are not painted with that fine rufous orange on the throat and breaſt, from which by a little ſtraining the red-breaſts derive their name. It tinges a few of the feathers by the end of Auguſt; and before the end of September the birds have all the ſame plumage, and can no longer be diſtinguiſhed †. About this time they prepare for their departure; but they never

* Belon, *Nat. des Oiſeaux*, p. 348.

† "It is improperly called red-breaſt (*rouge-gorge*); for what we think to be red on the breaſt is orange, which riſes from both ſides below the bill, which is ſlender, taper, and black, and from below the two ſpaces of the eyes, and extends from the under part of the throat to the ſtomach." *Id. ibid.*

 gather

gather in flocks: they journey singly, one after another; and, when all other birds associate together, the red-breasts still retain their solitary habits. They fly during the day from bush to bush; but probably they rise higher in the night, and make more progress; at least it has happened to fowlers in a forest which was full of these birds in the evening, and promised excellent sport next morning, that they were all gone before the dawn *.

The departure not being marked, or, to use the expression, not being proclaimed among the red-breasts, as among the other birds at that season collected in flocks, many stay behind; and these are either the young and inexperienced, or some which can derive support from the slender resources of the winter. In that season they visit our habitations, and seek the warmest and most sheltered situations †; and if any one still continues

* "I remember one year to have spread my nets for the red-breasts in April; and as they were very plenty, I continued the sport three days with equal success: on the fourth the sun rose brighter than ever, and the day was very mild; I expected a large capture, but they had beat their march during my absence; all were gone, and I did not take a single bird."

† "This bird being very delicate, and averse to excess of every kind, whether of heat or of cold, it retires in summer to the dense forests or the uplands, where it enjoys coolness and verdure; in winter it approaches dwellings, and is

tinues in the wood, it becomes the companion of the faggot-maker, cheriſhes itſelf at his fire, pecks at his bread, and flutters the whole day round him, chirping its ſlender pip. But when the cold grows more ſevere, and thick ſnow covers the ground, it approaches our houſes, and taps on the window with its bill, as if to entreat an aſylum, which is cheerfully granted*; and it repays the favour by the moſt amiable familiarity, gathering the crumbs from the table †, diſtinguiſhing affectionately the people of the houſe, and aſſuming a warble, not ſo rich, indeed, as that of the ſpring, but more delicate, and retained through all the rigours of the ſeaſon, to hail each day the kindneſs of its

is ſeen among the buſhes and in the gardens, eſpecially where the ſun ſtrikes, which aſpects it carefully ſeeks." Olina, *Uccelleria*, p. 16.

* Willughby, *Ornithol.* p. 160.

† "In a Carthuſian monaſtery at Bugey, I ſaw red-breaſts in the monks' cells, which they had been conſtrained to enter, after wandering ſome days in the cloiſters. In three or four days, they were naturaliſed to ſuch degree that they would come to eat on the table. They were well reconciled to the fare of the monaſtery; and paſſed thus the whole winter, protected from cold and hunger, without ſhewing the leaſt deſire to eſcape. But on the approach of ſpring they felt new deſires; they tapped on the window with their bill; and after they recovered their liberty, they returned not till next winter." *Note of M. Hebert.*

landlord,

landlord, and the ſweetneſs of its retreat*. There it remains tranquil, till the returning ſpring awakens new deſires, and invites to other pleaſures: it now becomes uneaſy, and impatient to recover its liberty.

During this tranſient ſtate of domeſtication, the red-breaſt eats almoſt every ſort of food; it collects indifferently crumbs of bread, bits of fleſh, or millet ſeeds. Olina aſſerts too generally, that, when taken from the neſt, or caught in the woods, it ſhould be fed with the ſame paſte as the nightingale †: its appetite, we ſee, is not ſo delicate. Thoſe which are allowed to fly freely about rooms, occaſion but ſlight inconvenience; for their excrements are ſmall and dry. The author of the Ædonology ‡ pretends, that the red-breaſts may be taught to ſpeak; and this prejudice muſt be of an ancient date, ſince the ſame thing occurs in Porphyry §: but the fact is quite improbable, for the tongue of the bird is forked. Belon, who never heard it except in autumn, at which time it has only the little chirp, and not its full, impaſſioned

* I ſaw, at the houſe of one of my friends, a red-breaſt that had been afforded an aſylum in the depth of winter, come to alight on his deſk while he was writing. It ſung whole hours with a feeble warble, ſweet and melodious.

† P. 16. ‡ P. 93. § Lib. iii. *De Abſtin. Animal.*

ſong

ſong of love, yet boaſts of the charms of its voice, and compares it to that of the nightingale. From his own account, he appears to have taken the red-breaſt and the red-ſtart for the ſame bird; but afterwards he diſtinguiſhed them, both by their habits and by their colours*. Thoſe of the red-breaſt are very ſimple: a coat of the ſame brown as the back of the throſtle covers all the upper ſide of its body and its head; its ſtomach and belly are white; the orange rufous of the breaſt is leſs vivid in the female than in the male; their eyes are black, large, and even expreſſive, and their aſpect mild; the bill is ſlender and delicate, as in all birds that live chiefly on inſects; the *tarſus*, which is very ſmall, is of a light brown, and ſo is the upper ſide of the toes, though they are of a pale yellow below. When full grown, the bird is five inches nine lines in length, and its alar extent eight inches; the inteſtinal tube is about nine inches long; the gizzard, which is muſcular, is preceded by a dilatation of the *œſophagus*; the *cæcum* is very ſmall, and ſometimes entirely wanting. In autumn, the red-breaſts are very fat; and their fleſh is more delicate than that

* "The red-ſtart appears in the ſpring in towns and villages, and neſtles in holes, when the red-breaſt has retired into the woods." Belon, *Nat. des Oiſeaux*, p. 348.

of the throſtle, but has ſomething of its flavour, both feeding on the ſame fruits, particularly thoſe of the ſervice-tree. [A]

[A] Specific character of the Red-breaſt, *Motacilla-Rubecula*: " It is gray; its throat and breaſt ferruginous." Its egg is whitiſh, with reddiſh ſpots. The robins ſeem to continue in England the whole year.

No. 126

THE BLUE-THROATED WARBLER.

THE

BLUE-THROAT.

Motacilla Suecica, *Linn. Gmel. & Frif.*
Sylvia Suecica, *Lath. Ind.*
Cyanecula, *Briff.*
Ruticilla Wegflecklin, *Ray & Will.*
The Blue-throated Red-ftart, *Edw.*
The Blue-throated Warbler, *Penn. & Lath.**

IN its fhape, its fize, and its general appearance, this bird is an exact copy of the red-breaft: the only difference is, that its throat is of a brilliant azure blue, while that of the other is orange red. Even in marking the difcrimination between thefe birds, nature feems to point at their analogy; for beneath this blue fpace, we may perceive a black fafh and an orange red zone which bounds the top of the breaft: this orange tint again appears on the firft half of the lateral quills of the tail; from the corner of the bill, a ftreak of rufty white paffes over the eye; and in general the colours, though darker, are the fame with thofe of the red-breaft. They fhare alfo the fame habits, only their haunts are different: the red-breaft

* In German, *Wegflecklin*, or *Blau-Kehlein:* in Swedifh, *Carls-Vogel.*

 lives

lives in the heart of the woods; but the blue-throat frequents their ſkirts, and ſeeks marſhes, wet meadows, and places that abound with willows and reeds: and with the ſame ſolitary inſtinct as the red-breaſt, it ſeems alike diſpoſed to be familiar with man; for, after the ſummer months are ſpent in its ſequeſtered retreats, it viſits the gardens, avenues, and hedges, before its departure, and ventures ſo near that it may be ſhot with the trunk.

Like the red-breaſts, they never keep in flocks, and ſeldom more than two are ſeen together. In the end of ſummer, ſays Lottinger, the blue-throats alight in the fields that are ſown with large ſorts of grain. Friſch mentions fields of peaſe as their favourite haunt, and even pretends that they breed among theſe; but their neſt is more commonly found in the willows, the oſiers, and other buſhes which grow in wet ſituations: it is formed of herbs interwoven at the origin of the branches or boughs.

In the love ſeaſon, the male mounts perpendicular to a little height in the air, chanting as he riſes; he whirls round, and drops back on his bough as cheerfully as the petty-chaps, of which the blue-throat ſeems to have ſome habits. He alſo ſings in the night; and, according to Friſch, his warble is very ſweet. Hermann, on the contrary, informs us, that it is not at all pleaſant *. This contradiction is owing, perhaps, to the different times at which theſe obſervers heard the bird; for as great difference would have been found between the ordinary chirp of our red-breaſt, and its mellow, tender ſong in the ſpring, or its pretty warble which cheers the bright days of autumn.

The blue-throat is as fond of bathing as the red-breaſt, and oftener haunts the margins of water. It lives on worms and other inſects; and in the ſeaſon of its paſſage it eats elder-berries. It may be ſeen among the fens, ſeeking its food on the ground, and running briſkly, and cocking its tail, eſpecially the male, when he hears the real or imitated voice of the female.

The young ones are of a blackiſh brown, and have not aſſumed the blue tint of the throat; the males have only ſome brown feathers amidſt the white of the throat and breaſt. The female never has the whole of the blue ſpace on the throat; ſhe is only marked with a creſcent or a bar below the neck: and on this difference, and on the figure of Edwards, who has given

* Doctor and Profeſſor of Phyſic and of Natural Hiſtory at Straſburg, who has been ſo obliging as to communicate ſome facts relating to the natural hiſtory of theſe birds.

the female only, Briſſon makes a ſecond ſpecies of his *Gibraltar blue-throat*, from which place the female ſeems to have come.

Among the adult males, ſome have the whole neck blue; and theſe are probably old ones, eſpecially as the reſt of the colours and the red zone of the breaſt appear deeper in theſe individuals. The others, which are more numerous, have a ſpot like a half collar, of a fine white, which Friſch compares to the luſtre of poliſhed ſilver*; and, from this character, the bird-catchers at Brandenburg have applied to the blue-throat the name of *mirror-bird*.

Theſe rich colours tarniſh and fade in confinement, and even after the firſt moulting. The blue-throats are caught with the nooſe, like the nightingales, and with the ſame bait. In the ſeaſon when theſe birds are fat, they are much ſought after, as are all the ſmall birds which have delicate fleſh; but they are rare, and even unknown in moſt of our provinces. They are ſeen at the time of their paſſage in low parts of the Voſges near Saneburg, according to Lottinger. But another obſerver aſſures us, that they never penetrate into the heart of thoſe mountains on the ſouth. They are more frequent in Alſace; and though they are ſpread

* Linnæus is probably miſtaken when he deſcribes this colour as a dull, yellowiſh white. *Fauna Suecica.*

through Germany, and even as far as Pruſſia, they are not very common in any part, and the ſpecies ſeems to be much leſs numerous than that of the red-breaſts. Yet it inhabits a wide extent: from the epithet which Barrere applies to it *, we may infer that it is well known among the Pyrenees. We ſee, from the denomination of Briſſon's *pretended* ſecond ſpecies, that this bird is found even at Gibraltar. It is alſo ſeen in Provence, where the people call it *blue ruſſet-bottom*. And from the name which Linnæus beſtows, we might ſuppoſe that it is a native of Sweden †; but this proves only that the blue-throat viſits the northern regions. It leaves them in autumn, and travels into milder climates in queſt of food; and this habit, or rather this neceſſity, is common to the blue-throat, and all the birds which feed on inſects, or on ſoft fruits. [A]

[A] Specific character of the Blue-throat, *Motacilla Suecica*: "Its breaſt is ferruginous, with a blue ſtripe; its tail quills duſky, and ferruginous near the baſe." It is found in the northern parts of Ruſſia and Siberia, but unknown in England.

* *Motacilla Pyrenaica.* † *Motacilla Suecica.*

FOREIGN BIRD,

WHICH IS RELATED TO THE RED-BREAST AND BLUE-THROAT.

THE BLUE RED-BREAST OF NORTH AMERICA.

Motacilla-Sialis, *Linn. & Gmel.*
Sylvia-Sialis, *Lath. Ind.*
Rubecula Carolinensis Cærulea, *Briss. & Klein.*
The Blue Red-breast, *Edw. & Catesb.*
The Blue Warbler, *Penn. & Lath. Syn.*

OUR red-breast is too weak, and its flight too short, for our supposing it to have crossed the Atlantic Ocean; and, as it cannot support extreme cold, it could not penetrate into America by the northern straits. But nature has produced its representative in those vast regions, which is found from Virginia and Carolina as far as Louisiana and the Bermuda islands. Catesby has given the first description of it, and Edwards has finished it; and both agree that it is closely related to the red-

red-breaſt of Europe*. It is rather larger than the red-breaſt, being ſix inches three lines in length, and its alar extent ten inches eight lines. Cateſby remarks, that it flies ſwiftly, and that its wings are long; the head, the upper ſide of the body, of the tail, and of the wings, is of an exceeding fine blue, except the tip of the wing, which is brown; the throat and breaſt are of a bright yellow ruſt colour; the belly is white. In ſome ſubjects, as in the one delineated by Cateſby, the blue tint of the head covers the throat alſo: in others, as in that of Edwards, the whole fore part of the body, as far as the bill, is covered with rufous. In the female, the colours are duller; the blue is mixed with blackiſh; the ſmall quills of the wing, blackiſh, and fringed with white. The bird is of a mild diſpoſition, and feeds only on inſects: it neſtles in the holes of trees—a difference in the mode of life ſuggeſted perhaps by the climate, where the numerous reptiles compel the birds to place the neſts beyond their reach. Cateſby tells us, that the blue warbler is very common through the whole of North America. That naturaliſt and

* "Mr. Cateſby has called this bird *Rubecula Americana;* which is a proper name enough, ſince both his bird and mine are certainly of that genus, of which the robin-red-breaſt is a ſpecies." *Edwards.*

and Edwards are the only ones who have noticed this bird; and Klein merely copies their defcriptions. [A]

[A] Specific character of the *Motacilla-Sialis*: "Above, it is blue: below, entirely red; its belly white." It breeds in hollow trees, and is very harmlefs and familiar.

THE

No. 127

FIG. 1 THE STONE CHAT. FIG 2. THE WHEAT-EAR.

THE

STONE-CHAT.

Le Traquet, *Buff.*
Motacilla-Rubicola, *Linn. Gmel. Kram.*
Sylvia-Rubicola, *Lath. Ind.*
Rubetra, *Aldrov. Johnst.* & *Briss.*
Sylvia Lutea, capite nigro, *Klein.*
Œnanthe tertia, *Ray, Will.* & *Sibb.*
The Stone-smich, Stone-chatter, or Moor-titling, *Will. Penn. Alb.**

THIS sprightly, active bird is never at rest. Fluttering from bush to bush, it alights only for a few seconds; and its wings are still spread, and ready to waft it to some other spot. It rises in the air by small springs, and falls back whirling round. This continual motion has been compared to that of a mill-clack (*traquet d'un moulin*); and hence, according to Belon †, the origin of its name.

* In Greek, Βάτις: in Italian, *Barada*, *Occhio di Bue* (ox-eye): in German, *Stein-schmeizer*. [Linnæus very absurdly gives the stone-chat the name *Rubicola*; which, bating a slight variation in the orthography, is absolutely the same with *Rubecula*, which was appropriated to the robin. T.]

† *Nat. des Oiseaux*, p. 360.

Though

Though the ſtone-chat flies low, and never mounts to the ſummit of trees, it ſits always on the tops of the buſhes, or on the moſt projecting branches of the hedges and ſhrubs, or on the heads of Turkey wheat growing in the fields, and on the talleſt props in the vineyards. It delights moſt in dry grounds, heaths, and upland meadows; and there it often utters its ſlender cry *wiſtratra*, in a low, ſmothered tone*. If it finds a detached ſtem or a ſtake in theſe meadows, it ſeldom fails to perch on it, which makes it eaſily be caught: a lime-twig placed on a ſtick is ſufficient for the purpoſe; and even the children are well acquainted with this ſport.

From this habit of flying from buſh to buſh, and on thorns and brambles, Belon, who found this bird at Crete and in Greece, as well as in our provinces, applies to it the name Βάτις or *bramble-bird*, which Ariſtotle mentions once only †, obſerving that it feeds on worms. Gaza tranſlates Βάτις by *rubetra*, which all the naturaliſts have referred to the ſtone-chat, eſpecially as the epithet *rubetra* implies that the bird is reddiſh, and the bay colour of the breaſt of the ſtone-chat is conſpicuous. It ſpreads, growing more dilute, under the belly: the ground co-

* Willughby's *Ornithology*, p. 170.

† *Hiſt. Anim.* lib. viii. 3.

lour

lour of the back is a fine black, which is clouded with brown ſcales; and the ſame diſpoſition of tints extends to the crown of the head *, where the black, however, predominates; the throat is of a pure black, only daſhed with a few white waves and riſes under the eyes. A white ſpot on the ſide of the neck is bounded by the black of the throat, and the bay colour of the breaſt; the quills of the wings and tail are blackiſh, fringed with brown or light ruſty; there is a broad white line on the wing near the body, and the rump is of the ſame caſt. All theſe tints are bolder and more intenſe in the old males than in the young ones; the tail is ſquare, and ſomewhat ſpread; the bill is ſlender, and ſeven lines in length; the head is pretty round, and the body compact; the legs are black, ſlender, and ten lines in length; the alar extent is ſeven inches and a half, and the total length four inches ten lines. In the female, the breaſt is of a dirty ruſty colour, which, mixing with the brown of the head and the upper ſide of the body, obtains a blackiſh caſt on the wings, and melts into the whitiſh under the belly and the throat; which makes the plumage of the female appear dull, diſcoloured, and much leſs diſtinct than in the male.

* Belon.

The

The ſtone-chat builds its neſt in waſte lands, at the foot of buſhes, under their roots, or beneath the cover of a ſtone, and enters it by ſtealth, as if afraid of being diſcovered: accordingly * the neſt is difficult to be found †. It breeds about the end of March, and lays ‡ five or ſix eggs, of a blueiſh green, with light rufous ſpots, which appear obſcure, but are moſt frequent at the big end. The parents feed their young with worms and inſects, which they

* "The *blackfoot* (ſtone-chat) makes its neſt in concealed ſpots. I found one plaſtered againſt a rock, two feet from the ground, in which were five young, covered with a black down. This neſt was concealed by a holm-oak, and the parents were not afraid that the cattle ſhould approach it; but they ſcreamed much on the neighbouring trees, when I went near." *Note communicated by the Marquis de Piolenc.*

† "They make their neſt ſo nicely, and frequent it ſo ſecretly, that it is very difficult to be found. It has a great many young, which it feeds with living animals." Belon, *Nat. des Oiſeaux*, p. 360.——" The neſt of the ſtone-chat is very difficult to diſcover; for the windings which the bird takes in entering or leaving it, render the ſearch almoſt always fruitleſs. It never enters but through ſome neighbouring buſhes; and when it iſſues forth, it flies likewiſe a ſhort way among the buſhes. A perſon might imagine, when he ſees the bird dart briſkly into a buſh with a worm or other inſect in its bill, that its neſt was there; yet in vain would he look for it at that ſpot."

Note communicated by the Sieur Trecourt.

‡ Neſt found at Montbard, the 30th March.

continually

continually carry to them: and their anxiety ſeems redoubled when theſe begin to fly; they invite them back, rally them, and ſcream inceſſantly *wiſtratra*; and for ſeveral days they ſtill bring them food.

The ſtone-chat is very ſolitary, and is always ſeen alone, except when the ſeaſon of love gives it a companion *. Its diſpoſition is wild, and its inſtinct dull and intractable: if it be ſprightly in the ſtate of freedom, it is as heavy † and inactive when reduced to domeſtication: it can be taught nothing, and it is even difficult to be raiſed ‡. In the fields it ſuffers one to come very near it, and flies to a ſhort diſtance, without ſeeming to notice the fowler: it appears, therefore, to have neither ſenſe enough to love us, nor ſhun us.

* Belon and Aldrovandus.

† "The ſtone-chat is penſive: having opened the cage to one of theſe birds, in a garden, amidſt buſhes, and in a hot ſun, it quickly flew to the open door, where it looked round it more than a minute before it took flight: its diffidence was ſo great as to ſuſpend its love of liberty."

Note communicated by M. Hebert.

‡ "The ſtone-chats are wild, and difficult to raiſe. Thoſe which I had, had a heavy look: ſometimes they took ſudden ſtarts; but theſe were momentary, and they ſoon relapſed into their habitual lethargy. They hopped, from time to time, upon ſomething elevated, ſhaking their wings and tail, and venting at intervals their cry *trac, trac*."

Note communicated by M. de Querhoënt.

The

The ſtone-chats are very fat in the proper ſeaſon; and, for the delicacy of their fleſh, they may be compared to the epicurean warblers. However, they live only on inſects, and their bill ſeems fit for nothing but grain. Belon and Aldrovandus aſſert, that they are not birds of paſſage: this may be true with reſpect to Greece and Italy; but it is certain, that, in the northern provinces of France, they ſhun the cold weather and the failure of the inſects, for they depart in September.

Some refer to this ſpecies the bird called in Provence the *fourmeiron* or *anter* *, becauſe it feeds chiefly on ants. The *anter* is ſolitary, and haunts only the ruins and rubbiſh of buildings: in cold weather, it ſits on the chimney-tops, as if to warm itſelf †. This feature would

* "The *fourmeiron* places itſelf at the hole of an ant-hill, ſo as to block it up completely; and the ants, preſſing to get out, entangle themſelves among its feathers: it then takes wing, and alights on ſome even ground, where it ſhakes its proviſion from its plumage; its table is ſpread, and it feaſts on its game. Itſelf is good eating."

Note of M. Guys, of Marſeilles.

† According to Meſſ. Guys and de Piolenc. But the latter, while he aſcribes this habit to the fourmeiron, judges it foreign to the ſtone-chats. "I have never heard (ſays he) that they love warmth; and I believe that I have ſeen them remove from the fires made in the fields to burn turf, which ſhews that they diſlike ſmoke." See the article of the Red-ſtart.

induce

induce us to range it rather with the red-ſtart than with the ſtone-chat, which conſtantly ſhuns towns and dwellings *.

In England, and particularly among the mountains of Derbyſhire, there is a bird which Briſſon terms the *Engliſh ſtone-chat*. Ray, who calls it the *coldfinch* †, ſays, that it is peculiar to that iſland: Edwards gives accurate figures of the male and female; and Klein mentions it by the name *variegated winged nightingale*. In fact, the white which marks not only the great coverts, but alſo the half of the ſmall quills neareſt the body, makes on the wing of this bird

* It is ſeen commonly in all places, but it never comes to the hedges of villages or towns. *Belon*, 360.

† Muſcicapa Atricapilla, *Linn. Gmel.* & *Kram.*
Rubetra Anglicana, *Briſſ.*
Curruca tergo nigro, *Friſch.*
The Pied Fly-catcher, *Penn.* & *Lath.*

Specific character: " It is black; the under ſide, the ſpot on its front, and the ſpangle on its wings, white; its lateral tail quills, white exteriorly."——Thus deſcribed by Briſſon: " Above, black; below, white; its rump variegated with black and white; a bright white ſpot on its forehead, and a white one on the wings; the leſſer wing quills, white without, black within; the outermoſt, white exteriorly (*in the male*); above, dirty greeniſh brown; below, white; a yellowiſh white ſpot on the wings; the exterior wing quills, yellowiſh white; the inner, blackiſh; the bird a ſpot much broader than in the common ſtone-chat. Beſides, the white colour covers all the fore and under part of the body, and makes a ſpot on the face; and the black ſpreads over the upper ſide of the body to the rump, which is barred with black and white: the quills of the tail are black; the outermoſt, white exteriorly; and the great quills of the wing, brown. All that is black in the male, is, in the female, of a dirty greeniſh brown; the reſt is alſo white: in both, the bill and legs are black.—This bird is of the ſize of the ordinary ſtone-chat; and though it ſeems peculiar to England, and even to the hills of Derbyſhire*, it muſt migrate, for ſometimes it is ſeen in Brie.

The ſpecies of the ſtone-chat is ſpread from England † and Scotland ‡, as far as Italy and Greece. It is very common in many parts of France. Nature ſeems to have exhibited it in the South, under various forms. We ſhall treat of theſe foreign ſtone-chats, after deſcribing a

the tail quills, blackiſh; the outermoſt, fringed exteriorly with white."

* It is found ſometimes alſo in Shropſhire and Flintſhire. T.

† Willughby.

‡ Sibbald.

ſpecies which much reſembles it, and inhabits the ſame climates. [A]

[A] Specific character of the Stone-chat, *Motacilla-Rubicola*: " It is gray; below, tawny; a white ſtripe on its throat; its ſtraps black." Its egg is blueiſh green, with dull rufous ſpots. Whether this ſpecies entirely leaves England in winter, is uncertain.

THE

WHIN-CHAT.

Le Tarier.
Motacilla-Rubetra, *Linn. Gmel. Kram. Frif. Mull.* & *Brun.*
Sylvia-Rubetra, *Lath. Ind.*
Rubetra Major, feu Rubicola, *Briff.*
Œnanthe fecunda, *Ray* & *Will.*
Sylvia Petrarum, *Klein* *.

THE fpecies of the whin-chat, though clofely related to that of the ftone-chat, ought yet to be feparated from it, fince both inhabit the fame places without intermixing; as in Lorraine †, where they are very common, but live

* In German, *Flugen-Stakerle*, *Flugen-Stakerlin*, *Todten-Vogel*: in Silefian, *Noeffel-Fincke*.

† "There is another little bird, of the bulk of the ftone-chat, different from all other birds, in its habits, its flight, and in its mode of living and of building its neft; which the inhabitants of Lorraine call *tarier*. It lives in bufhes, like the ftone-chat; and has a flender bill, calculated for flies and worms: its nails, its legs, and its feet, are black; but the reft of its body refembles that of the mountain-finch, for it has a whitifh fpot acrofs the wing, like the finch and the ftone-chat: however, its bill and its mode of life will not permit it to be ranged with the mountain-finches.—The male has fpots on his back, and round his neck;

live diſtinct. Their habits are different, as well as their plumage. The whin-chat ſeldom perches; it is generally on the ſurface, among mole-hills, in fallow grounds, and in the high paſture lands beſide the woods: the ſtone-chat, on the contrary, ſits always in the buſhes, or on vine-props. The whin-chat is alſo larger than the ſtone-chat, its length being five inches and three lines: the colours are nearly the ſame, only differently diſtributed; in the former, the upper ſide of the body is marked with brighter tints, each wing has a double ſpot, and a white line extends from the corner of the bill to the back of the head*; a black ſpace lies under the eye, and covers the cheeks, but does not ſpread, as in the ſtone-chat, under the throat, which is of a light bay; this bay gradually ſoftens, but may be diſtinguiſhed on the white ground on all the fore part of the body; the rump is of the ſame white colour, though more intenſe, and ſpeckled with black; all the upper ſide of the body as far as the crown of the head, is of a black, grounded and ſpotted with brown; the ſmall quills and the great coverts are black. Willughby

neck; his head reſembles that of the mountain-finch; the extremities of the wings and of the tail are ſomewhat purple, as in that finch; but the bird is leſs ſpotted: ſome pretend that it is a ſpecies of ſtone-chat."

Belon, *Hiſt. des Oiſeaux*, p. 361.

* Willughby.

 ſays,

ſays, that the end of the tail is white: we have obſerved, on the contrary, that the quills are white in their firſt half from the root. But that naturaliſt himſelf found varieties in that part of the plumage of the whin-chat; and mentions his having ſometimes ſeen the two middle quills of the tail black, with a rufous border; and, at other times, with the ſame edging on a white ground. - The female differs from the male, in its colours being paler, and the ſpots on the wings being not ſo apparent: it lays four or five eggs, of a dirty white, dotted with black *.— The whin-chat builds its neſt like the ſtone-chat; it arrives and departs with it, participates of its ſolitary inſtinct, and ſeems even more ſhy and wild; it prefers the mountainous tracts, and in ſome places it derives its name from that circumſtance: thus the Bologneſe fowlers call it *montanello*; and the appellations which Klein † and Geſner ‡ give to it, mark its fondneſs for ſolitude in bleak and rugged ſpots. The ſpecies is not ſo numerous as that of the ſtone-chat §: it feeds alſo on worms, flies, and inſects:

* Mr. Latham ſays, that the egg of the whin-chat is uniform ſky blue. T.

† Sylvia Petrarum.

‡ Paſſerculi genus ſolitarium.

§ Belon.

laſtly,

laſtly, it is very fat about the end of ſummer, and it is then ſcarcely inferior in delicacy to the ortolan. [A]

[A] Specific character of the Whin-chat, *Motacilla Rubetra*: " It is blackiſh, its eyebrows white, a white ſpot on its wings, and its throat and breaſt yellowiſh." It is a bird of paſſage in Great Britain.

FOREIGN BIRDS,

WHICH ARE RELATED TO THE STONE-CHAT AND WHIN-CHAT.

I.

THE SENEGAL STONE-CHAT OR WHIN-CHAT.

Motacilla Fervida, *Gmel.*
Sylvia Fervida, *Lath. Ind.*
The Sultry Warbler, *Lath. Syn.*

THIS bird is of the ſize of the whin-chat, and ſeems cloſer allied to it than to the ſtone-chat. Like the former, it has the double white ſpot on each wing, and no black on the throat; but it has not the black ſpace under the eye; nor are its great wing coverts black, being only ſpotted black on a brown ground: in other reſpects, the colours are nearly the ſame as in the ſtone-chat or whin-chat, only they are more vivid over all the upper ſide of the body; the brown of the back is of a lighter rufous, and the black daſhes are there better defined. This pleaſing variety prevails from the

the crown of the head to the coverts of the tail, the middle quills of the wings are edged with rufous, the great ones with white, but more lightly; they are all blackiſh. But if the colours be more diſtinct on the upper ſide of this bird than in the common ſtone-chat, they are, on the contrary, duller below, only the breaſt is ſlightly tinged with a fulvous red between the white of the throat and that of the belly.—This bird was brought from Senegal by Adanſon. [A]

[A] Specific character of the *Motacilla Fervida*: "It is brown; the margin of its quills rufous; below, ochry-white; two white ſpots on the wings; the tail black."

II.

THE STONE-CHAT FROM THE ISLAND OF LUCON.

Motacilla-Caprata, *Linn. & Gmel.*
Sylvia-Caprata, *Lath. Ind.*
Rubetra Lucionenſis, *Briſſ.*
The Luzonian Warbler, *Lath. Syn.*

THIS ſtone-chat is not quite ſo large as that of Europe, but it is rounder and ſtronger; its bill is thicker, and its legs not ſo ſlender. It is entirely of a dark brown, except a broad white bar

bar on the coverts of the wing, and a little of a dull white under the belly. The female might, from its colours, be taken for a bird of quite a different ſpecies; a dun rufous covers all the under ſide of her body and the rump; the ſame colour ſhews itſelf on the head through the waves of a browner caſt, which grow deeper on the wings and tail, and become of a very dull ruſty brown.—The cock and hen were ſent from the iſland of Luçon, where, Briſſon ſays, they are called *maria-capra*. [A]

[A] Specific character of the *Motacilla-Caprata*: "It is black; its rump, its vent, the ſpot on its wing coverts, white."

III.

STONE-CHAT OF THE PHILIPPINES.

Motacilla Fulicata, *Linn. & Gmel.*
Sylvia Fulicata, *Lath. Ind.*
Rubetra Philippenſis, *Briſſ.*
The Sooty Warbler, *Lath. Syn.*

IT is of a ſtill deeper black than the male of the preceding. It is larger, being near ſix inches; and its tail longer than in any of the ſtone-chats; its bill and legs are alſo ſtronger: the only

only interruption of its plumage, which is black with violet reflexions, is the white ſpot on the wing. [A]

[A] Specific character: "It is dark violet, its vent cheſnut, the ſpot on its coverts white."

IV.

THE GREAT STONE-CHAT OF THE PHILIPPINES.

Motacilla Philippenſis, *Gmel.*
Sylvia Philippenſis, *Lath. Ind.*
Rubetra Philippenſis major, *Briſſ.*
The Philippine Warbler, *Lath. Syn.*

THIS ſtone-chat is larger than the preceding, being ſomewhat more than ſix inches in length; its head and throat are white, daſhed with ſome ſpots of reddiſh and yellowiſh: a broad brick-coloured collar decorates the neck; and, below it, a ſcarf of a blueiſh black covers the breaſt, and extends to the back, forming a ſhort cope, with two large white ſpots thrown on the ſhoulders; the reſt of the upper ſide of the body is black, with violet reflexions to the end of the tail; this black is intercepted on the wing by two

two ſmall white bars, the one on the outer edge near the ſhoulder, the other at the extremity of the great coverts; the belly and ſtomach are of the ſame reddiſh white with the head and throat; the bill, which is ſeven lines long, and the legs, which are thick and ſtout, are of a ruſt colour. Briſſon ſays that the legs are black, and perhaps their colour varies: the wings reach to the end of the tail, which is different from what obtains in all other ſtone-chats, where the wings ſcarcely extend to the middle of the tail.

V.

THE FITERT OF MADAGASCAR STONE-CHAT.

Motacilla-Sybilla, *Linn. & Gmel.*
Sylvia-Sybilla, *Lath. Ind.*
Rubetra Madagaſcarenſis, *Briſſ.*
The Sybil Warbler, *Lath. Syn.*

BRISSON has given a deſcription of this bird, which we find to be very exact, on comparing it with a ſpecimen ſent to the King's cabinet. He tells us, that it is called *fitert* at Madagaſcar, and that its ſong is pleaſant; this ſeems to ſhew, that it is different from our ſtone-chats,

 which

which have only a disagreeable chirp: they agree, however, in several prominent characters. It is rather larger than the European stone-chat, being five inches four lines in length; the throat, the head, and all the upper side of the body to the tail, are black, only the back and shoulders have some rusty waves; the fore part of the neck, the stomach, and the belly, are white; the breast is rufous; the white colour of the neck is set off by the black of the throat and the rufous of the breast, and forms a sort of collar; the great coverts of each wing which are next the body are white, which makes a white spot on the wing; a little white also terminates the quills of the wing on their inside, and augments in proportion as they are nearer the body. [A]

[A] Specific character of the *Motacilla-Sybilla*: "It is blackish, below white, the breast rufous, a white spot on the wings."—Thus described by Brisson: "Above, black; the feathers fringed with rusty at their tips; the breast rufous; a bright spot on the wings; the quills of the tail black."

VI. THE

VI.

THE GREAT STONE-CHAT.

Motacilla Magna, *Gmel.*
Sylvia Magna, *Lath. Ind.*
The Dark Warbler, *Lath. Syn.*

THIS ſtone-chat well deſerves the epithet of *great*. It is ſeven inches and a quarter from the point of the bill to the extremity of the tail; and ſix inches and an half from the point of the bill to the end of the nails; the bill is an inch long, and not ſcalloped; the tail is about two inches, and is a little forked; the wings, when cloſed, cover the half of it; the *tarſus* is eleven lines; the mid toe ſeven, the hind one ſeven, and its nail the ſtrongeſt of them all. Commerſon has left an account of this bird, but makes no mention of the country where he ſaw it; however, the deſcription we give of it is ſufficient to enable travellers to recogniſe it. The chief colour of its plumage is brown; the head is variegated with two brown tints; a light brown covers the upper ſide of the neck and body; the throat is mixed with brown and whitiſh: the breaſt is brown; and this is alſo the colour of the coverts of the wings, and of the outer edge of the quills, their

their inſide being divided by rufous and brown; and this brown appears again at the extremity of the quills of the tail, and covers the half of the middle ones, the reſt being rufous; and the outſide of the two outer feathers is white; the under ſide of the body is ruſty.

VII.

THE STONE-CHAT OF THE CAPE OF GOOD HOPE.

Sylvia Sperata, *Lath. Ind.*

M. DE ROSENEUVETZ ſaw, at the Cape of Good Hope, a ſtone-chat which has never been deſcribed by naturaliſts. It is ſix inches long; the bill black, and ſeven lines in length, ſcalloped near the tip; the legs black; the *tarſus* an inch long: all the upper ſide of the body, including the neck and head, of a very brown green; all the under ſide of the body, gray, with ſome tints of rufous; the rump is rufous; the quills and the coverts of the wings are brown, with a lighter border of the ſame colour; the tail is twenty-two lines in length, and the wings reach to its middle; it is ſomewhat forked: the two middle

middle quills are of a blackiſh brown; the two ſide ones are marked obliquely with brown on a fulvous ground, and this is the more remarkable in proportion as they are exterior. Another ſpecimen, of the ſame ſize, brought alſo from the Cape of Good Hope by M. de Roſeneuvetz, and depoſited in the Royal Cabinet, is perhaps only the female. The whole upper ſide of the body is merely of a blackiſh brown, the throat whitiſh, and the breaſt rufous. We are unacquainted with the habits of theſe birds; yet theſe are what alone form the intereſting part of the hiſtory of animated beings. But how often have we been forced to regret, that we could only deſcribe the appearance, and not delineate the character! All theſe incomplete facts ought however to be collected, and placed beſide the immenſe range of obſervations; as the navigators inſert in their charts the lands ſeen at diſtance.

VIII. THE

VIII.

THE SPECTACLE WARBLER.

Le Clignot, ou Traquet a Lunette, *Buff.*
Motacilla Perſpicilla, *Gmel.*

A CIRCLE of yellowiſh ſkin folded quite round the eyes, and reſembling ſpectacles, is a ſingular character which diſtinguiſhes this bird. Commerſon found it on the river *de la Plata*, near Montevideo, and the appellations* which he gives to it, allude to this uncommon conformation of the exterior of the eyes. It is of the ſize of a goldfinch, but thicker in the body: its head is round, and the crown raiſed: its plumage is of a fine black, except the white ſpot on the wing, which makes it reſemble the ſtone-chats; this ſpot ſpreads broad over the middle of the five firſt quills, and terminates in a point about the end of the ſixth, ſeventh, and eighth. In ſome ſubjects, there is alſo white on the lower coverts of the tail; in others, they are black as the reſt of the plumage: the wing reaches only to the middle of the tail, which

* *Perſpicillarius, Nictitarius, Lichenops,* Clignot.

is two inches long, ſquare when it is cloſed, and forming almoſt an equilateral triangle when it is ſpread; it conſiſts of eight equal quills: the bill is ſtraight, ſlender, yellowiſh at the upper part, and ſlightly bent or hooked at the end: the tongue is membranous, like a double-pointed arrow: the eyes are round; the iris yellow, and the pupil blueiſh. The ſingular membrane which encircles the eyes, is probably nothing but the ſkin of the eye-lid naked and more expanded than ordinary, and conſequently broad enough to make ſeveral folds. Such at leaſt is the idea which Commerſon ſuggeſts by comparing it to the wrinkled lichen, and telling us that the two portions of this membrane, which is fringed at the edges, meet when the bird ſhuts its eyes. We may alſo obſerve the *membrana nictitans*, which riſes from the inner *canthus*. The legs and toes are ſlender and black; the hind toe is the thickeſt, and is as long as thoſe before, though it has only one joint, and its nail is the ſtrongeſt. Could this bird be bred ſeparate from the reſt of its kind, and exiled in the middle of the new continent? It is at leaſt the only one in America that is known to be related to the chats; but the analogies which it bears to them are not ſo ſtriking as the character which diſtinguiſhes it, and which nature has impreſſed as the ſtamp of thoſe foreign lands which it inhabits. [A]

[A] Specific character of the *Motacilla Perſpicillata*: "It is black, the coverts of the wings marked with a white ſpot; the tail equal; the orbits naked, yellowiſh, and wrinkled."

THE

WHEAT-EAR.

Le Motteux, *Buff.* *
Motacilla-Œnanthe, *Linn. & Gmel.*
Sylvia-Œnanthe, *Lath. Ind.*
Œnanthe, *Gesner, Johnst. &c.*
Vitiflora, *Briss. & Klein.*
Culo Bianco, *Zinn.*
The Wheat-ear, Fallow Smich, or White-tail, *Penn. Alb. Edw. & Lath.*†

THIS bird is common in the country, and is continually among the clods in new-tilled fields, and hence its name in French ‡. It follows the furrow traced by the plough, and searches for worms on which it feeds. When it is scared away, it never mounts high, but

* The old name in French was *Vitrec*; the vulgar one at present *Cul Blanc*.

† In Greek, Οιναυθη, according to Belon, from οινη a vine, and ανθος a flower: the Latin *Vitiflora* is a translation of the Greek: in Italian, *Culo Bianco, Fornarola, Petragnola*: in German, *Stein-Schwaker, Stein-Schnapperl*: in Swedish, *Stensguetta*: in Norwegian, *Steen-Dolp, Steen-Gylpe*.

‡ *Motteux*, from *motte* a clod.

skims

ſkims along the ſurface with a ſhort rapid flight, and in its retreat it ſhews the white of the hind part of its body, by which it is eaſily diſtinguiſhed in the air from all other birds, and hence its vulgar appellation among fowlers, *cul-blanc* *. It is alſo pretty frequent in fallow grounds, where it flies from ſtone to ſtone, and ſeems to ſhun the hedges and buſhes, on which it does not perch near ſo often as on clods.

It is larger than the whin-chat, and taller, on legs which are black and ſlender; the belly is white, and ſo are the upper and under coverts of the tail, and nearly the half of its quills, of which the tips are black; they ſpread when it flies, and expoſe the white for which it is remarkable: the wing in the male is black, with ſome fringes of ruſty white: the back is a fine aſh gray or blueiſh gray, which extends to the white ground; a white ſpot riſes at the corner of the bill, bends under the eye, and ſtretches beyond the ear; a white ſtripe bounds the face, and paſſes over the eyes. The female has neither this ſpot nor this ſtripe; its plumage is marked with a ruſty gray wherever that of the male is aſh gray; its wing is more brown than black, and broad fringed as far as below the

* White Arſe.

belly; and on the whole it resembles as much, or more, the hen whin-chat than its proper male. The young resemble the parent birds exactly at the age of three weeks, the time at which they fly.

The bill of the wheat-ear is slender at the tip and broad at the base, which enables it to seize and swallow the insects, on which it runs, or rather darts, rapidly by a succession of short hops *. It is always on the ground; and if it be put up, it only removes from one clod to another, flies always exceedingly low, and never enters the woods, nor perches higher than the hedges or small bushes. When seated it wags its tail, and chirps with a dull sound *titreû, titreû* †; and, as often as it flies, it seems to pronounce distinctly, with a stronger voice, the words *far-far, far-far*: it repeats these two cries with a degree of precipitancy.

It breeds under the tufts and clods in newly ploughed fields, and under stones in fallow grounds, near quarries, in old rabbit burrows ‡, or in the naked stone walls which are used for fences in hilly countries. Its nest is constructed with care: it is composed of moss or tender grass, and lined with feathers

* Belon, *Nat. des Oiseaux*, p. 352.

† Hence perhaps its old French name *vitrec*, or *titrée*.

‡ Willughby.

or

or wool; it is diſtinguiſhed by a ſort of covert placed above it, and ſtuck to the ſtone or clod under which the whole is formed. It lays generally five or ſix eggs, of a light blueiſh white, with a circle at the large end of a duller blue. A female, which was caught on the eggs, had loſt all the feathers from the middle of the ſtomach, as uſual in the caſe of vigilant ſitters. The male is attentive to his mate, and during the time of incubation he brings ants and flies: he watches near the neſt; and when he obſerves one paſſing, he runs or flies before, and endeavours to draw notice till the perſon has got to a ſufficient diſtance, and then he returns by a circuit to the neſt.

The young ones are ſeen as early as the middle of May; for theſe birds have returned to our provinces as early as the firſt fine days in March *. But froſts often ſurpriſe them after their arrival, and numbers periſh; as happened in Lorraine in 1767 †. There are many of them in that province, eſpecially in the mountainous part: they are equally common in Burgundy and Bugey; but they are hardly ſeen in Brie, except towards the end of the ſummer ‡. In general they prefer high countries, upland plains, and arid tracts. Great

* Lottinger. † Id. ‡ Hebert.

numbers are caught by the English shepherds in the downs of Sussex about the beginning of autumn, at which time they are plump and of delicate flavour. Willughby describes the method of catching them: they cut up a long strip of turf, and invert it on the furrow, so as to leave only a narrow track, in which they place snares made of horse hair. The birds are incited by a double motive; to procure food in the new-turned earth, and to conceal themselves under the sod. The appearance of a hawk, or even the shadow of a cloud, will drive them for shelter into those traps *.

They all return in August and September, and no more are seen after that month. They journey in small bands; and in general they are of a solitary disposition, and no society exists among them but that of the male and female. Their wings are large †, and though

* Mr. Pennant tells us, that in the district of *Eastbourn* in Sussex, one thousand eight hundred and forty dozens of wheat-ears are at an average caught annually, which are sold commonly for sixpence a dozen. The reason why these birds are so numerous in that neighbourhood, is said to be the abundance of a certain species of fly, on which they feed. T.

† Brisson says, that the first of the wing quills is extremely short: but the feather which he takes to be the first of the great quills is only the first of the great coverts, inserted under the first quill, and not at the side of it.

among us they make little uſe of their power of flying, they probably exert it in their migrations. They muſt have once done ſo: for they are among the few birds which are common to Europe and the ſouth of Aſia; ſince they are found in Bengal, and inhabit the extent of Europe, from Italy to Sweden.

The appellations * which the wheat-ear receives in different parts of France, allude to its habits of living on the ground and in the holes; of ſitting on the clods, and appearing to ſtrike them with its tail. Its Engliſh names refer to its frequenting both fallow and tilled grounds, and to the whiteneſs of its rump. But the Greek term *œnanthe*, which naturaliſts have, from a conjecture of Belon, agreed to apply to it, ſeems not ſo characteriſtic or ſo proper as the preceding. The mere analogy of the word *œnanthe* to *vitiflora*, and the reſemblance of this to the old French name *vitrec*, led Belon to form this opinion; for he does not explain why it was called *vine-flower* (οινανθη). It alſo arrives before the blowing of the vine, and continues long after the bloſſom is dropped; and it has therefore no connexion with the flower of the vine. Ariſtotle deſcribes it only

* *Motteux, Tourne-motte, Briſe-motte, & Terraſſon*; *i. e.* clodder, turn-clod, break-clod, and earth-thrower.

as appearing and disappearing at the same time with the cuckoo *.

Brisson reckons five species of wheat-ears:

I. THE WHEAT-EAR.

II. THE GRAY WHEAT-EAR, which he discriminates from the first only by that epithet, though that is equally gray. Its difference, according to Linnæus, who makes it a variety, is, that the plumage, which is of a pale colour in both, is marked with small whitish waves. Brisson adds another slight distinction in the breast feathers, which are, he says, sprinkled with little gray spots; and in those of the tail, of which the two middle ones have no white, though the rest are white three fourths of their length. But the minute details of the various tints of the plumage would easily transform the same individual into several species; we have only to describe it nearer or farther from the season of moulting †. To examine the productions

* *Hist. Anim.* lib. ix. 49. Pliny says the same of the disappearance of the *œnanthe*, lib. x. 29. From this passage, Father Hardouin infers that the *œnanthe* is not the wheat-ear, but a nocturnal bird.

† Young wheat-ears, taken the 20th of May, had the upper part of their body mottled with rusty and brown: the feathers

productions of nature in this way is to lose sight of her design; it is to mistake the sportive superficial touches of her pencil, for the deep permanent strokes with which she has engraved the characters of animals.

III. The third species of Brisson is THE CINEREOUS WHEAT-EAR*; but the differences which he marks are too slight to discriminate them, especially since the epithet *cinereous* agrees as well with the common wheat-ear, of which this is only a variety. Thus the three pretended species are reduced to one. But the fourth and fifth species of Brisson are more decidedly distinguished, viz. *The Rusty White-Tail*, and *The Rufous White-Tail*.

THE RUSTY WHEAT-EAR †, which is Brisson's

feathers of the rump are whitish, striped lightly with black; the throat and the under side of the body rufous, dotted with black: all this livery is cast the first year.

* "Above cinereous white, mixed with gray brown; below white; the rump gray brown; the lower part of the neck light tawny; the forehead bright white; a black spot below the eyes: of the two middle quills of the tail their first half is white, and the other blackish; the lateral ones white, terminated with blackish; the three outermost on both sides fringed with whitish at the tips." *Brisson.*

† "White; the top, the upper part of the back, and the breast, dilute tawny; a black bar on the eyes; the two middle tail quills black, fringed with black on both sides near the tip."

ſon's fourth ſpecies, is rather leſs than the common wheat-ear, being only ſix inches and three lines in length: the head, the fore part of the body, and the breaſt are whitiſh, mixed with a little rufous; the belly and the rump are of a lighter white; the upper ſide of the neck and back is light ruſty. It might be readily taken for the female of the common wheat-ear, if ſome individuals had not the character of the male, the black ſtripe on the cheek between the bill and the ear; ſo that this would ſeem to be a permanent variety. It is found in Lorraine near the mountains; but it is not ſo frequent as the ordinary ſpecies. It alſo inhabits the vicinity of Bologna in Italy; and Aldrovandus calls it *ſtrapazzino*. Briſſon tells us that it occurs in Languedoc, and that at Nimes it is termed *reynauby*.

The fifth ſpecies of Briſſon * is THE RUFOUS WHEAT-EAR. Both male and female have

* Motacilla Stapazina, *Gmel.*
Sylvia Stapazina, *Lath. Ind.*
Specific character: "It is ferruginous; its wings brown; the ſpace about its eyes and its tail black; its outermoſt tail quill white at the edge." Thus deſcribed by Briſſon: "Yellowiſh rufous; its rump and lower belly white (the cheeks and throat black *in the male;* a black bar on the eyes *in the female*); the two middle tail quills black; the lateral ones white, fringed with black."

been

been defcribed by Edwards, who received them from Gibraltar. One of them had not only the black ftripe between the bill and the ear, but its throat was entirely of the fame colour: a character that was wanting in the other, whofe throat was white, and the tints paler; the back, the neck, and the crown of the head were of a yellow rufous; the breaft, the top of the belly, and the fides, were of a diluter yellow; the lower belly and the rump, white; the tail white, fringed with black, except the two middle quills, which are entirely black; thofe of the wing are blackifh, and their great coverts edged with light brown. This bird is nearly the fize of the common wheat-ear. Aldrovandus, Willughby, and Ray fpeak of it under the name of *œnanthe altera*. We may regard it as a fpecies clofely related to the common wheat-ear, but much lefs frequent in the temperate countries. [A]

[A] Specific character of the Wheat-ear, *Motacilla-Oenanthe:* "Its back is hoary; its front white; a black ftripe on the eyes." In England the wheat-ears arrive between March and May, and retire in September.

FOREIGN BIRDS

WHICH ARE RELATED TO THE WHEAT-EAR.

I.

THE GREAT WHEAT-EAR, OR WHITE TAIL OF THE CAPE OF GOOD HOPE.

Motacilla Hottentotta, *Gmel.*
Sylvia Hottentotta, *Lath. Ind.*
The Cape Wheat-ear, *Lath. Syn.*

M. De Roſeneuvetz ſent us this bird, which has not been deſcribed by any naturaliſt. It is eight inches long; its bill ten lines, its tail thirteen, and the *tarſus* fourteen. It is much larger than the European kind: the upper ſide of the head is ſlightly variegated with two browns, whoſe tints melt into each other; the reſt of the upper ſide of the body is fulvous brown as far as the rump, where there is a tranſverſe bar of light fulvous; the breaſt is variegated, like the head, with two ſhades of brown, which are confuſed and indiſtinct; the throat is dirty white, tinged with brown; the higher

higher part of the belly and the flanks are fulvous; the lower belly is dirty white, and the inferior coverts of the tail light fulvous; but the superior ones are white, and so are the quills as far as their middle: the rest is black, terminated with dirty white, except the two middle ones, which are entirely black, and tipped with fulvous; the wings are of a brown cast, edged slightly with light fulvcus on the great quills, and more slightly on the middle quills and on the coverts.

II.

THE GREENISH BROWN WHEAT-EAR.

Motacilla Aurantia, *Gmel.*
Sylvia Aurantia, *Lath. Ind.*
The Orange-breasted Wheat-ear, *Lath. Syn.*

THIS species was also brought from the Cape of Good Hope by M. de Roseneuvetz. It is smaller than the preceding, being only six inches long; the upper side of its head and body is variegated with black, brown, and greenish brown: these colours also mark distinctly the coverts of the wings; but the great coverts

coverts of the wings, and thoſe of the tail, are white: the throat is dirty white; there is alſo a mixture of that colour and of black on the fore part of the neck: the breaſt is tinged with orange, which grows dilute below the belly; the inferior coverts of the tail are entirely white; the quills are blackiſh brown, and the lateral ones are tipped with white. This bird has, ſtill more than the preceding, all the characters of the common wheat-ear, and we can ſcarcely doubt that their habits are nearly the ſame.

III.

THE SENEGAL WHEAT-EAR.

Motacilla Leucorhoa, *Gmel.*
Sylvia Leucorhoa, *Lath. Ind.*
The Rufous Wheat-ear, *Lath. Syn.*

IT is rather larger than the European ſpecies, and reſembles the female exactly; only the back has a little more of the reddiſh caſt. [A]

[A] Specific character: " Duſky rufous; below ochry white; the rump, the coverts, and the baſe of the tail, white."

THE

THE

WAGTAILS.

THE White Wagtail *(Lavandiere)* has often been confounded with the other kinds *(Bergeronettes)*: but the former commonly haunts the ſides of pools, and the others frequent the meadows, and follow the flocks. All of them flutter often in the fields round the huſbandman, and attend the plough to pick up the worms that crawl in the freſh-turned ſoil. At other ſeaſons, the flies which moleſt the cattle, and all the inſects which ſwarm on the margin of ſtagnant water, are their food. The wagtails are real *fly-catchers*, if we regard only their manner of life: but they differ from theſe birds, becauſe they do not watch their prey from trees, and hunt it: they only ſearch on the ground. They form a ſmall family of birds with a delicate bill, tall and ſlender legs, and a long tail, which they vibrate inceſſantly: and, from this habit, they have been termed *Motacilla* by the Romans, and received their various names in the provinces of France.

THE

WHITE WAGTAIL.

La Lavandiere, *Buff.*
Motacilla Alba, *Linn. Gmel. Kram. Frif. Mull. Will*
Motacilla, *Briff.*
Sylvia Pectore Nigro, *Klein.*
Coda Tremula, *Zinn.*
Ballarina, *Olina.*
Bachſtelzen, *Gunth. & Wirſ.**

BELON, and, before him, Turner, applied to this bird the Greek name κνιπολογος, rendered into Latin by *culicilega* or *gnat-gatherer*; and that appellation would ſuit the wagtail,

* In Latin, *Motacilla*: in Italian, *Ballarina*, *Coda-tremola*, *Codin-zinzola*, *Cutretola*, *Bovarina*: in Catalonian, *Cugumela*, *Marllenga*: in Portugueſe, *Aveloa*: in German, *Wyſſe Waſſer Steltze* (white water-ſtilts), *Bach Steltze* (brook-ſtilts), *Weiſſe und Schwartze Bach Steltze* (white and black brook-ſtilts), *Wege-Stertze* (weigh-tail), *Kloſter-Stertze* (cloiſter-tail): in Flemiſh, *Quick-Stertz*: in Swediſh, *Aerla*, *Saedes-Aerla*: in the dialect of Oſtrobothnia, *Waeſtraeckia*: in Norwegian, *Erle*, *Lin-Erle*: in Daniſh, *Vip-Stiert*, *Havre Sœer*: in Poliſh, *Pliſka*, *Trzeſiogonek Bialy*.

Near Montpelier it is called *Enguane-Paſtre*: in Guyenne, *Peringleo*: in Saintonge, *Batajoſſe*: in Gaſcony, *Battiquoüe*: in Picardy, *Semeur*: at Nantes, and in Orleanois, *Bergeronette*,

No. 128

FIG.1. THE WHITE WATERWAGTAIL. FIG.2. THE SHEPHERDESS.

wagtail, though I am confident the κνιπολογος was quite a different bird.

Ariſtotle (lib. viii. c. 3.) ſpeaks of two wood-peckers (δρυοκολαπτας)*, of the golden oriole (κολιος, or *galgulus*), as lodging in trees, which they ſtrike with their bill. To theſe muſt be joined, he ſays, the little gnat-gatherer (κνιπολογος) †, which is ſpotted with gray, and hardly ſo large as the goldfinch, and with a feeble voice. Scaliger properly obſerves that a *lignipeta* (ξυλοκοπων) ‡, or pecker of trees, cannot be a wagtail. A gray ſpeckled plumage is different from that of the wagtail, which is interſected with great bars, and mottled with white and black ſpots. Nor are the characters of ſize and of feeble voice applicable to the wagtail, of which we cannot diſcover either the name or the deſcription in the Greek authors; though all theſe properties belong to the common creeper §.

Bergeronette, or *Vachette*: in Lorraine, *Hoche-Queue*: in Burgundy, *Croſſe-Queue*, *Branle-Queue*: in Bugey, *Damette*: and in the other provinces of France, *Lavandiere*.

* From δρυς an oak, and κολαπτω to beat.

† Perhaps from κωνωψ a gnat, and λεγω to gather.

‡ From ξυλον wood, and κοπτω to cut.

§ Turner himſelf was in the end convinced, that the κνιπολογος was a kind of wood-pecker; and Aldrovandus thinks that Ariſtotle meant by that name a creeper.

The white wagtail is ſcarcely larger than the ordinary titmouſe, though its long tail ſeems to add to its ſize, ſo that its whole length is ſeven inches: the tail itſelf is three inches and an half, which the bird expands and diſplays while it flies. With this large oar it directs and balances its motions: it whirls, it darts, and ſports in the air; and when it alights, it briſkly wags it upwards and downwards, at intervals of five or ſix ſhakes.

Theſe birds run nimbly with little haſty ſteps on the ſandy brinks; they even venture with their long legs to the depth of a few lines in the thin ſheet of water that ſpreads over the ſhelving margin: but they oftener flutter about mill-dams, and ſit on ſtones. They viſit the waſher-women, and hover about them the whole day, approaching familiarly, and picking up the crumbs that are thrown to them; and, by the jerking of their tail, ſeem to imitate the action of cleanſing linen; from which habit they have been called in French *lavandiere (waſher)* *.

The plumage of the white wagtail conſiſts of mottles and large ſpots of black and white: the belly is white: the tail conſiſts of twelve quills, of which the ten middle ones are black,

* Belon.—In England they are likewiſe called ſometimes *diſh-waſhers*. T.

and

and the two ſide ones white to near their origin: the wing reaches only the third of their length; the quills of the wings are blackiſh and white gray. Belon obſerves that, with regard to its wings, the wagtail has ſome relation to the aquatic birds*. The upper ſide of the head is covered with a black cap, which deſcends to the nape of the neck; a white half-maſk conceals the face, ſurrounds the eye, and, falling on the ſides of the neck, bounds the black of the throat, which is marked with a broad horſe-ſhoe rounded on the breaſt. Many ſubjects have only a zone or ſemi-circle at the top of the breaſt, and their throat is white; and the back, which is of a ſlate gray in others, is of a brown gray in theſe, which ſeem to form a variety †, though they are mixed and confounded

* "It has a particular mark by which it reſembles the ſhore birds: this is, that the laſt feathers of the wings, joining the body, are as long as the firſt of the anterior ones; which obtains likewiſe in all other birds that live on flies and earth worms, the plovers and the lapwings."

Belon, *Nat. des Oiſeaux*, p. 349.

† "The lead colour varies in this kind of birds, ſome being more cinereous, others blacker." *Willughby*. Albin ſays the ſame, vol. i. p. 43. Some obſervers ſeem to attribute this difference to that of age, and aſſert that moſt of the wagtails are white on their return in ſpring, and aſſume black in the courſe of the ſeaſon. Belon ſeems to be of this opinion: "The young wagtails in their ſixth month," ſays

he,

founded with the ſpecies; for the difference between the male and female is, that in the latter the crown of the head is brown; but in the former it is black *.

The white wagtail returns into our provinces about the end of March. It breeds on the ground under ſome roots, or below a graſs tuft in lands not in tillage: but ofteneſt by the edge of waters, beneath a hollow bank, or under the ſtakes of wood that are driven along the ſides of rivers. Their neſt conſiſts of dry herbs and ſmall roots, ſometimes intermixed with moſs; the whole looſely compoſed, and lined with feathers or hair. They commonly have four or five white eggs, ſprinkled with brown ſpots; and only make a ſingle hatch, unleſs the firſt fails. The parents defend their young courageouſly: they flutter and dive before their enemy to draw him aſide; and if he carries off the neſt, they follow him, flying above his head, and conſtantly whirling round, calling on their young with doleful cries. They are alſo attentive to the cleanlineſs of their family, and throw out the excrements, or even

he, "are of another colour than thoſe an year old, and which have caſt their firſt plumage."

* "In this ſpecies the female differs from the male in having the ſpot on its head, not black, but gray." *Olina*. "The female has an aſh-coloured top." *Schwenckfeld.*

remove

remove them to a certain diſtance. They alſo diſperſe the bits of paper and ſtraws which have been laid to mark their neſt *. After the young are able to fly, the parents continue to feed and train them for three weeks or a month: they gorge greedily the inſects and ants' eggs that are brought to them †. Theſe birds are always remarked to eat uncommonly quickly, without ſeeming to allow time for ſwallowing. They collect the worms on the ground; they purſue and catch the flies in the air, and theſe are often the objects of their whirling. Their flight is waving, and conſiſts of jerks and ſprings. They aſſiſt their motion by vibrating

* "I obſerved wagtails that built in a hole of a wall waſhed by the river: they were at pains to clean their neſt, and carry the excrements more than thirty paces off. A piece of white paper happened to reſt on the ſtake that propped the wall by the water edge: this ſeemed offenſive to the wagtails; and I ſaw them, one after another, make fruitleſs efforts to remove it. It was too heavy, and I therefore took it away; but left in its place little ſtrips of paper equally white. They would not ſuffer theſe to remain; but carried them to the ſame diſtance as the dung of their young, being deceived by the ſimilarity of colour. I repeated this experiment ſeveral times."—*Note communicated by M. Hebert.*

† "I put eggs of large ants in a place where the wagtails reſorted: they took fifteen or ſixteen each time, till their throat was filled, and then carried them to their young."—*Note of the ſame obſerver.*

their tail horizontally; a motion different from that on the ground, which is performed perpendicularly. The wagtails utter frequently, especially while on the wing, a ſmall, ſhrill, redoubled cry, and in a clear tone, *gŭit-gŭit gŭigŭigŭit:* it is the note of rallying, for thoſe on the ground anſwer it. But the cry is louder, and oftener repeated, when they have juſt eſcaped the talons of the hawk*. They are not ſo much afraid of men or other animals; for when they are fired at, they do not fly far, but return to alight at a ſhort diſtance from the fowler. Some are caught along with the larks, by means of the net and mirror †; and it appears from Olina's account, that in Italy they are particularly fowled for about the middle of October.

Autumn is the time when they are moſt numerous in the country ‡. That ſeaſon, which

* Olina.

† This ſport laſts from four in the afternoon till the duſk of the evening: the perſons place themſelves by the margin of water, and attract the wagtails with a decoy bird of the ſame ſpecies; or, if that cannot be had, with ſome other ſmall bird.

‡ "In Brie, in Burgundy, in Bugey, and in moſt of our provinces, prodigious numbers are ſeen at certain times near inhabited places; in the fields, following the flocks: whence it appears that they are birds of paſſage."—*Note of M. Hebert.*

8 collects

collects them together, seems to inspire them with cheerfulness: they multiply their sports; they hover in the air, fall in the fields, pursue and call upon each other. They come forward in numbers on the roofs of mills, and in hamlets near water, and appear to hold discourse together by their little broken and repeated cries; we might fancy that they interrogate each other, and, for a certain time, reply in their turns, till the general acclamation of the assembly marks their resolution or consent to remove to some other spot. Now it is that they have the little soft warble with a low voice, which scarcely exceeds a murmur*; and from this circumstance, probably, Belon has applied to them the Italian name *susurada* (from *susurrus*, a whisper). This gentle breathing is prompted by autumn, and by the pleasures of society, to which these birds seem much attached.

About the end of autumn, the wagtails form into larger bodies. In the evening they descend among the willows and osiers, by sides of streams and rivers, where they call those which pass, and together make a noisy wrangling till dusk. In the clear mornings of October they fly sometimes very high, and vociferate incessantly

* Belon.

ſantly to each other. Then is the time when they migrate into other climates *. M. de Maillet ſays that, in this ſeaſon, prodigious numbers of them drop in Egypt, and that the people dry them in the ſand to preſerve them for eating †. M. Adanſon mentions that they are ſeen in winter at Senegal, with the ſwallows and quails, but only during that ſeaſon.

The white wagtail is common through the whole of Europe, as far as Sweden, and is found too, as we have juſt noticed, in Africa and in Aſia. The one which M. Sonnerat brought to us from the Philippines is the ſame with that of Europe. That brought from the Cape of Good Hope by Commerſon differed not from the variety delineated *Pl. Enl. fig.* 2. *No.* 652, except that the white of its throat did not riſe on its head, nor ſo high on the ſides of the neck, and that the coverts of its wings are leſs varied, and do not form the two tranſverſe white lines.

* "In the north of England it appears not in winter, and rarely even in the ſouth." *Willughby.* "The white wagtails depart in autumn." *Geſner.*

† "From Cairo to the ſea, all along the Nile, but chiefly near dwellings, are ſeen a great number of wagtails (*bergeronettes* ou *lavandieres*) of the blueiſh gray ſpecies, with a black half collar, ſhaped like a horſe-ſhoe. I could not be informed whether theſe birds remained the whole year in Egypt."—*Note ſent from Cairo by M. Soniini.*

But

But is not Olina miſtaken in aſſerting that the white wagtail is not ſeen in Italy, unleſs in autumn and winter? and is it likely that this bird ſpends the winter in that country, when it puſhes its migrations ſo far into much hotter climates*? [A]

[A] Specific character of the White Wagtail, *Motacilla Alba*: "Its breaſt is black; its two lateral tail quills divided obliquely with white."

* *Uccelleria*, p. 51.

THE

BERGERONETTES, OR BERGERETTES.

THE GRAY BERGERONETTE.

FIRST SPECIES.

Motacilla Cinerea, *Gmel. & Briſſ.*
The Cinereous Wagtail, *Lath.*

WE have ſeen that the *lavandiere*, or white wagtail, conſiſts of a ſingle ſpecies, that admits only of a ſlight variety; but the family of the *bergeronettes* includes three very diſtinct ſpecies, and all of them live in our fields without aſſociating or breeding together. Not to interfere with the received names, we ſhall denominate them *the gray bergeronette*, *the ſpring bergeronette*, and *the yellow bergeronette*; and we ſhall, in a ſeparate article, notice the foreign birds related to theſe.

The ſort of attachment which theſe birds ſhew to flocks; their habits of following them in in the meadows, and of fluttering amidst cattle, while these are feeding*, and sometimes even alighting on the backs of cows and sheep; their familiarity with the herdsman, whom they attend with confidence and security, and give notice of the approach of the wolf, or of the bird of rapine: all these circumstances have procured them an appellation suited to this pastoral life†. The companion of innocent and peaceful men, the *bergeronette* displays that attachment to our species, which would unite to us most animals, were they not repulsed by our barbarity, and the apprehension of becoming our victims. In the *little shepherdess* love predominates above fear; no bird at liberty in the fields appears so tame ‡: it allows one to gain nearer and nearer it, and seems not to avoid the fowler §.

It feeds on flies during the summer months; but after the frosts have destroyed the winged insects, and confined the cattle to their stalls, they retire to the brooks, and there pass almost

* "When these birds follow the herds, they are the spies, or rather the sentinels, of the keeper; for they give notice when they descry a wolf, or a ravenous bird."—*Note communicated by M. Guys.*

† The word *bergeronette*, or *bergerette*, signifies a little shepherdess.

‡ Belon.

§ Salerne.

the

the whole of the ſevere ſeaſon. At leaſt, the moſt of them continue with us during the winter: the yellow *bergeronette* is more uniformly ſtationary: the gray is leſs common in that ſeaſon.

All the *bergeronettes* are ſmaller than the white wagtail, and their tail is proportionally longer. Belon was well acquainted only with the yellow one, and appears to indicate the gray *bergeronette* by the appellation of *another kind of lavandiere*.

The upper ſide of the gray *bergeronette* is gray, or cinereous; the under ſide of its body white, with a brown bar, or half collar, on the neck: the tail is blackiſh, with white on the outer quills: the great quills of the wings are brown; the others blackiſh, and fringed with white, like the coverts.

They build about the end of April, commonly on a willow near the ground, and ſheltered from rains. They breed twice a year. The ſecond hatch is late; for their neſts are found even in September; which could never happen to a family of birds that migrate, and are obliged to educate their young before the winter. However, thoſe of the firſt hatch, and the pairs which have more diligently diſcharged their office, ſpread through the fields in the months of July and Auguſt; whereas the white

wagtails

wagtails ſeldom flock, except when they migrate about the end of September and in October*.

The *bergeronette*, which is conſtitutionally the friend of man, will not become his ſlave, and it dies in the cage. It loves ſociety, and cannot bear cloſe confinement; but, if left looſe in a room during winter, it will ſurvive, and will catch flies, and pick up the crumbs of bread†. Sometimes it alights on board ſhips, becomes familiar with the ſailors, continues with them in the voyage, and never leaves them till their arrival at the port‡. But ſuch facts may perhaps be aſcribed to the white wagtail, which roams more than the *bergeronette*, and which, in paſſing the ſeas, is apt to loſe its way. [A]

[A] Specific character of the Gray Bergeronette, *Motacilla Cinerea:* "It is cinereous gray, below white; a tawny ſtripe on the breaſt *(in the male)*; the tail black: the greateſt part of the two outermoſt tail quills is white." It is unknown in Britain.

* Belon.

† Geſner and Schwenckfeld.

‡ "On the 8th of June we were off the coaſts of Sicily, twelve or fifteen leagues from land. We caught on the veſſel a *bergeronette:* we ſet it at liberty, but it ſtill continued with us. Food and drink were ſet for it on one of the windows, to which it regularly came for its meal. It faithfully accompanied us till we were cloſe on the iſle of Candia: it quitted us when we had entered the port of Sonda."—*Note communicated by M. de Manoncour.*

THE

THE

SPRING BERGERONETTE.

SECOND SPECIES.

Motacilla Flava, *Linn. Gmel. Mull. Kràm. Friſ. Ray.*
Sylvia Flavia, *Klein.*
Motacilla Verna, *Briſſ.*
The Yellow Wagtail, *Penn. Will. Edw. & Lath.* *

THIS *bergeronette* is the firſt that is ſeen in the meadows and fields, where it neſtles among the green corn. Scarcely indeed does it diſappear in the winter, unleſs during the moſt ſevere colds: it commonly haunts, like the yellow kind, the ſides of brooks, and ſprings which never freeze. The epithets beſtowed on theſe birds ſeem improper; for the following ſpecies has leſs yellow than the preſent. That colour is diſtinct only on the rump and belly; but, in the ſpring *bergeronette*, all the upper and fore parts of the body are of a fine yellow; and there is a ſtreak of the ſame on the

* In German, *Bach-Steltze* (brook-ſtilts), *Gelbruſtige* (yellow-breaſt), *Irlin*, *Gelber Sticherling*, *Gelbe-Weyer-Bach Steltze* (yellow-weighing-brook-ſtilts).

wing,

wing, at the fringe of the middle coverts. All the mantle is of a dull olive, which alſo borders the eight quills of the tail, whoſe ground colour is blackiſh: the two outer ones are more than half white: thoſe of the wings are brown, with their outer edge whitiſh; and the third of thoſe neareſt the body reaches, when the wing is cloſed, as far as the longeſt of the great quills; a character which we have already noticed in the white wagtail. The head is cinereous; the crown tinged with olive: above the eye there is a line, which is white in the female, and yellow in the male; which is diſtinguiſhed alſo by blackiſh ſtreaks, more or leſs frequent, forming a creſcent under the throat, and alſo ſprinkled above the knees. When the male is in ſeaſon, he runs and turns round his female, briſtling up the feathers on his back in an odd ſort of way, but which undoubtedly expreſſes the fire of his paſſion. Their hatch is ſometimes late, but commonly productive. They breed often under the banks of rivulets, and ſometimes in the midſt of corn before harveſt*. They frequent, in autumn, the herds of cattle, like the other *bergeronettes*. The ſpecies is common in England, in France †, and ſeems to

* Willughby. Edwards.

† Edwards.

be ſpread through the whole of Europe, as far as Sweden *. We have found, in ſeveral ſubjects, the hind nail to be longer than the great fore toe; an obſervation which Edwards and Willughby had made before, and which contradicts the axiom of the nomenclators, who aſſume it as a generic character of theſe birds, that this nail and this toe are equal †. [A]

[A] Specific character of the *Motacilla Flava*: "Its breaſt and belly are yellow; its two lateral tail quills parted obliquely by white. Its egg is lead coloured, variegated with yellowiſh ſpots."

* Linnæus. † Briſſon.

THE

THE

YELLOW BERGERONETTE.

THIRD SPECIES.

Motacilla Boarula, *Gmel.*
Motacilla Flava, *Briss.*
The Yellow Wagtail, *Alb.*
The Gray Wagtail, *Edw. Penn. Will. & Lath.**

WHEN the white wagtails depart in autumn, the *bergeronettes* come near our dwellings, ſays Geſner, and appear even in the midſt of the villages. This habit belongs eſpecially to the yellow kind †: it then procures its ſubſiſtence beſide the margins of perennial ſprings, and ſhelters itſelf beneath the ſhelving banks of rivulets. It finds its ſituation ſo comfortable, that it even warbles in that torpid ſeaſon, unleſs the cold be exceſſive. This is a ſoft whiſpered ſong, like the autumnal notes of the white wagtail, and very different from the ſhrill cry which it utters in riſing into the air.

* In Italian, *Coda-tremola Gialla;* in German, *Kleine Bach Steltze* (little brook-ſtilts): in Poliſh, *Pliſka-Zolta.*

† Geſner, Aldrovandus, Olina.

In

In the ſpring, it removes to breed in the meadows, or ſometimes in the copſes beneath a root, and near running water: the neſt is placed on the ground, and built with dry herbs and moſs, well lined with feathers, hair, or wool, and cloſer interwoven than that of the white wagtail. It contains ſix, ſeven, or eight eggs, of a dirty white, ſpotted with yellowiſh. After the young are raiſed, and the meadows are mowed, the parents lead them among the herds of cattle.

Flies and gnats are then their food; for, when they haunt the ſides of ſtreams in winter, they ſubſiſt on worms, and alſo ſwallow little ſeeds. We found theſe, with fragments of caterpillars, and a ſmall ſtone, in the gizzard of a yellow *bergeronette*, caught in the end of December. The *æſophagus* was dilated before its inſertion: the gizzard was muſcular, and lined with a dry wrinkled membrane, which had no adheſion: the inteſtinal tube was ten inches long, and without any *cæcum*, or gall bladder: the tongue was fringed at the end, as in all the *bergeronettes*: the hind nail was the longeſt.

Of all the long tailed birds, the yellow *bergeronette* is moſt remarkable for that character *: its tail is near four inches, and its body is

* Edwards.

only

only three and an half: its alar extent is eight inches ten lines: its head is gray; its mantle, as far as the rump, deep olive, on a gray ground; its rump yellow; the under ſide of its tail of a brighter yellow; its belly and breaſt of a pale yellow in young ſubjects, ſuch as thoſe which Briſſon ſeems to have deſcribed; but in adults they are of a rich brilliant yellow *: the throat is white: a ſmall longitudinal whitiſh bar riſes at the bill, and paſſes over the eye: the plumage of the wings is of a brown gray, ſlightly fringed in ſome places with a white gray: there is ſome white at the origin of the middle quills, which forms a tranſverſe bar on the wing, when this is ſpread; alſo, the exterior edge of the three neareſt the body is pale yellow, and of theſe three the firſt is almoſt as long as the largeſt quill: the outermoſt of thoſe of the tail is entirely white, except a black hollow on the inſide: the next is white only within, and the third the ſame: the ſix others are blackiſh. Thoſe which have on the throat a black ſpot, bearing a white bar under the

* Edwards.—“There is a diſtinction in the bergerette between the male and the female; the male being very yellow under the belly, no bird more ſo.” *Belon.*

cheek, are the males*. According to Belon, their yellow tint is alſo much more vivid: the line of the eyebrows is equally yellow: and it is remarked that the colour of all theſe birds is more intenſe in winter after moulting.

Edwards deſcribes this bird under the name of *the gray water wagtail* †; and Geſner applies the epithets of *ſhake-tail*, *beat-ley*, which are equivalent to *lavandiere (waſher-woman)*. In fact, theſe *bergeronettes* frequent, no leſs often than the white wagtail, the brinks of water and pebbly brooks ‡; and, ſince they lodge in ſuch ſituations during the winter, their haunts are even more conſtant. However, the greater part of them migrate; for they are more numerous among the cattle in autumn, than beſide the ſprings and rivulets in winter §. Linnæus

* Willughby deſcribes only the female, and calls it the *gray wagtail*: and Albin, who gives two figures of this bird, only delineates the female twice; for neither of them has black on its throat.

† *Gleanings*. An inaccurate denomination which originated with Willughby, who owns that he deſcribed only the female.

‡ Willughby.

§ "In the month of Auguſt ſuch numbers are caught, that hundreds are brought to town, although at other ſeaſons they are rare, and cannot be got." *Belon*. Adanſon found

næus and Frisch take no notice of this species; whether because they confound it with the spring bergeronette, or because only one of these occurs in the north of Europe [A].

[A] Specific character of the *Motacilla Boarula*: "Above cinereous; below yellow; the whole of its first tail quill, and the inside of the second, white." It is frequent in England; breeds in the northern part of the island, and shifts in winter to the south.

The *Java bergeronette* of Brisson resembles much this third species. The differences are slight, or even vanish in comparing the descriptions; and we shall not hesitate to class them together *.

found the yellow bergeronette in Senegal. "In this isle (Goree) are many small water-birds, woodcocks of several kinds, larks, thrushes, sea partridges, and common wagtails, which are the ortolans of the country; being little pellets of fat, excellently flavoured." *Voyage to Senegal*, p. 169.

* "Above ash brown, inclining to olive; below yellow; the lower part of the neck and the breast dirty gray, with a mixture of yellow on the breast; the outermost tail quill white; the two next white on the inside, and at the tips." *Motacilla Javensis*. Brisson.

FOREIGN BIRDS

WHICH ARE RELATED TO THE BERGERONETTES.

I.

THE BERGERONETTE FROM THE CAPE OF GOOD HOPE.

Motacilla Capensis, *Linn. Gmel. & Briss.*
The Cape Wagtail, *Lath. Ind.*

THE foreign bergeronettes resemble so closely those of Europe, that we might readily suppose them to be derived from the same stock, and only modified by the influence of climate. The one from the Cape of Good Hope was brought by Sonnerat, and is the same which Brisson describes. A great brown mantle, which terminates in black on the tail, and its two edges, joined below the tail by a brown scarf, covers all the upper side of the body, which is as large as that of the white wagtail. All the under

under ſide of the body is dirty white; a ſmall line of the ſame colour interſects the brown hood on the head, and paſſes from the bill to the eye. Of the quills of the tail the eight middle ones are entirely black: the exterior on each ſide are broad-ſcalloped with white: the wing appears brown when cloſed; but, on ſpreading it, the half of its length is white. [A]

[A] Specific character of the *Motacilla Capenſis*: "It is brown; below whitiſh; a brown ſtripe on its breaſt; its eyebrows white; its lateral tail quills obliquely white."

II.

THE LITTLE BERGERONETTE FROM THE CAPE OF GOOD HOPE.

Motacilla Afra, *Gmel.*
The African Wagtail, *Lath.*

THERE are two characters which oblige us to ſeparate this bird from the preceding. 1. The ſize; this one being only five inches, of which the tail occupies two and an half. 2. The colour of the belly, which is entirely yellow, except the inferior coverts, which are white:

white: a ſmall black bar paſſes over the eye, and ſtretches beyond it: all the mantle is of a yellowiſh brown: the bill is broad at its baſe, and grows thinner at the middle, and more inflated at the tip: it is black, as are alſo the wings and the legs: the toes are very long; and Sonnerat, who brought it, obſerves that the hind nail is larger than the reſt: he obſerves too, that this ſpecies reſembles much the following, which he has alſo communicated, and which is perhaps the ſame, varied only by the difference of climate between the Cape and the Moluccas.

III.

THE BERGERONETTE OF THE ISLAND OF TIMOR.

Motacilla Flava, *Var. Lath. Ind.*
The Timor Wagtail, *Lath. Syn.*

AS in the preceding, its body is yellow: there is a ſtreak of the ſame colour on the eye: the upper ſide of its head and body is cinereous gray: the great coverts are tipped with white, and form a bar of the ſame colour on the wing,

wing, which is black, as well as the tail and the bill: the legs are pale red: the hind nail is twice as long as the reſt: the bill, as in the preceding, is at firſt broad, then thin, and afterwards ſwelled: the tail is twenty-ſeven lines, and exceeds the wings eighteen lines; and the bird ſhakes it continually, like the European wagtails.

IV.

The BERGERONETTE FROM MADRAS.

Motacilla Maderaſpatenſis, *Gmel.*
Motacilla Maderaſpatana, *Briſſ.*
The Pied Wagtail, *Lath.*

RAY firſt noticed this ſpecies, and from him Briſſon has drawn his deſcription *; but neither of them mentions the ſize. Its colours conſiſt of black and white: the head, the throat, the neck, and all the back, including the wings, are black: all the quills of the tail are white, except the two middle ones; theſe are black, and

* "Black *(male)*, cinereous *(female)*; belly white; a bright white longitudinal bar on the wings; the two middle tail quills black; the lateral ones black." *Briſſon.*

and rather ſhorter than the reſt, which makes the tail forked. the belly is white: the bill, the legs, and nails are black. Every part that is black in the male, is gray in the female.

THE

No. 129

FIG. 1. THE FIG-EATER FIG. 2. THE PITPAT.

THE

FIG-EATERS.

Les Figuiers, *Buff.*

THESE birds are of a genus approaching to that of the fig-peckers, and resemble these in their principal characters. Their bill is straight, slender, and very acute, with two small scallops near the extremity of the upper mandible; a property which they have in common with the tanagres, in which however the bill is much thicker and shorter. The nostrils of the fig-eaters are uncovered, which distinguishes them from the titmice: the angle of their hind nail is arched, which separates them from the larks; and therefore they must be ranged by themselves.

We are acquainted with five species of fig-eaters in the hot countries of the old continent, and twenty-nine in those of America: these differ from the former in the shape of the tail, which is regularly tapered in the species that inhabit the old continent, but notched at the end in the natives of America, and almost forked, the two middle quills being shorter than the others;

others; and that character is sufficient to decide to what continent they belong. We shall begin with those of the old.

THE

GREEN AND YELLOW FIG-EATER.

LE FIGUIER VERT & JAUNE, *Buff.*

FIRST SPECIES.

Motacilla Tiphia, *Linn. & Gmel.*
Sylvia Tiphia, *Lath. Ind.*
Ficedula Bengalensis, *Briss.*
The Green Indian Fly-catcher, *Edw.*
The Green Indian Warbler, *Lath. Syn.*

THIS bird is four inches and eight lines in length; its bill seven lines, its tail twenty lines, and its legs seven lines and an half: the head and all the upper side of the body are of an olive green; the under side of the body yellowish: the superior coverts of the wings are of a deep brown, with two transverse white bars: the quills of the wings and those of the tail are of the same green with the back: the bill, the legs, and the nails are blackish.

 Edwards

Edwards defcribes this bird as brought from Bengal, and terms it a *fly-catcher*, though its bill indicates a quite different genus. Linnæus is alfo miftaken in reckoning it a wagtail *(motacilla)*; for the tails of the fig-eaters are much fhorter. [A]

[A] Specific character of the *Motacilla Tiphia*: "It is green; below yellowifh; the wings black; two white bars."

THE

CHERIC.

SECOND SPECIES.

Motacilla Madagafcarienfis, *Gmel.*
Sylvia Madagafcarienfis, *Lath. Ind.*
Ficedula Madagafcarienfis Minor, *Briff.*
The White-eyed Warbler, *Lath. Syn.*

IN the ifland of Madagafcar this bird is known by the name *tcheric*: it was tranfported into the ifle of France, where it is called *white-eye*, on account of a fmall white membrane encircling its eyes. It is fmaller than the preceding, being only three inches and eight lines in length,

length, and its other dimenſions proportional: its head, the upper ſide of its neck, its back, and the ſuperior coverts of its wings, are of an olive green: its throat and the inferior coverts of its tail are yellow: the upper ſide of its body is whitiſh: the quills of the wings are of a light brown, and bordered with olive green on their outer margin; the two quills in the middle of the tail are of the ſame olive green with the upper ſide of the body: the other quills of the tail are brown, and edged with olive green: the bill is dun gray: the legs and nails are cinereous. The Viſcount Querhoënt, who obſerved this bird in the iſle of France, ſays that it is not timid, yet ſeldom viſits the ſettlements; that it flies in flocks, and feeds on inſects. [A]

[A] Specific character of the *Motacilla Madagaſcarienſis*: "It is olive brown: its head rufous; its throat white; its breaſt tawny; its belly rufous brown."

THE

THE

LITTLE SIMON.

THIRD SPECIES.

Sylvia Borbonica, *Lath. Ind.*
Ficedula Borbonica, *Briſſ.*
The Bourbon Warbler, *Lath. Syn.*

THIS bird is called the *Little Simon* in the iſland of Bourbon, though it is not a native of that place, and muſt have been tranſported thither; for we are informed by people of veracity, and particularly by Commerſon, that there exiſted no kind of quadrupeds or birds in the iſlands of Bourbon and France, when the Portugueſe firſt diſcovered them. Theſe iſlands appear to be the points of a continent which has been ſwallowed up, and almoſt their whole ſurface is covered with volcanic productions; ſo that at preſent they are ſtocked only with animals that have been carried to them.

This bird is exactly of the ſame ſize with the preceding: the upper ſide of its body is of a light ſlate colour; the under ſide white gray; the throat white; the great quills of the tail

 deep

deep brown, edged on one ſide with a ſlate colour: the bill is brown, acute, and ſlender; the legs gray, and the eyes black: the females, and even the young ones, have nearly the ſame plumage as the males. They are very numerous in every part of the iſland of Bourbon, where the Viſcount Querhoënt obſerved them. They uſually breed in September, and lay three or four eggs, probably ſeveral times in the courſe of the year. They build on ſingle trees, and even in orchards: the neſt is formed of dry herbs, and lined with hair: the eggs are blue. Theſe birds will allow a perſon to get very near them; they fly always in flocks, and feed on inſects and ſmall ſoft fruits. When they ſee a partridge running along the ground, a hare, or a cat, &c. they flutter round it, making a peculiar cry; and hence they direct the fowler to his prey. [A]

[A] Specific character of the *Sylvia Borbonica*, Lath.: "It is brown gray; below yellowiſh gray; the quills of the wings and of the tail edged with gray."

THE

BLUE FIG-EATER.

LE FIGUIER BLEU, *Buff.*

FOURTH SPECIES.

Motacilla Mauritiana, *Gmel.*
Sylvia Mauritiana, *Lath. Ind.*
The Maurice Warbler, *Lath. Syn.*

THIS ſpecies has not been noticed by any naturaliſt, and is probably a native of Madagaſcar. The male ſeems to differ in nothing from the female, except that its tail is a ſlight degree longer, and the upper ſide of its body has a tinge of blue mixed with the whitiſh. The head and all the upper ſide of the body are of a blueiſh cinereous: the quills of the wings and of the tail are blackiſh, edged with white: the bill and legs are blueiſh. [A]

[A] Specific character of the *Motacilla Mauritiana:* "It is blue gray; below white; the quills of the wings and tail black, edged with white."

THE

SENEGAL FIG-EATER.

FIFTH SPECIES.

WE conceive that the three birds delineated No. 582. *Pl. Enl.* are the ſame ſpecies; of which the ſpotted fig-eater is the male, and the two others only varieties ariſing from age or ſex. They are all very ſmall, but figure 1. is the leaſt *.

The ſpotted fig-eater †, No. 2, is ſcarcely four inches long, of which the tail occupies two: it is tapered, and the two middle quills are the longeſt: all theſe tail quills are brown, fringed with ruſty white; ſo are alſo the great quills of the wings. The plumage of the wings, and of the back and head, is black, edged with light rufous: the rump is deeper rufous, and the fore part of the body is white.

The two others differ from this, but reſem-

* Sylvia Rufigaſtra, *Lath.*

† Motacilla Undata, *Gmel.*
Sylvia Undata, *Lath. Ind.*
The Undated Warbler, *Lath. Syn.*

ble

ble each other. The fig-eater (figure 3) * has not its tail tapered: it is light brown, and proportionally ſhorter than the body: the upper ſide of the head and body is brown: the wing is blackiſh brown, fringed on the quills, and undated on the coverts with a ruſty brown: the fore part of the body is of a light yellow, and there is a little white under the eyes.

The fig-eater (figure 1) is ſmaller than the other two: all its plumage is nearly the ſame as that of figure 3, except the fore part of the body, which is not light yellow, but aurora red.

We have already ſeen that, in ſome ſpecies of the genus of fig-eaters, there are ſome individuals whoſe colours vary conſiderably.

We preſume likewiſe, that the three other birds of No. 584. *Pl. Enl.* are of the ſame identical ſpecies; of which the firſt appears to be the male †, and the two others varieties of age or ſex ‡; the third, particularly, ſeems to be a

* Motacilla Flaveſcens, *Gmel.*
Sylvia Flaveſcens, *Lath. Ind.*
The Citron-bellied Warbler, *Lath. Syn.*

† Motacilla Fuſcata, *Gmel.*
Sylvia Fuſcata, *Lath. Ind.*
The Duſky Warbler, *Lath. Syn.*

‡ Motacilla Subflava, *Gmel.*
Sylvia Subflava, *Lath. Ind.*
The Flaxen Warbler, *Lath. Syn.*

 female.

female. In all the three, the head and upper ſide of the body are brown; the under ſide gray, with a flaxen tint of various extent and intenſity: the bill is brown, and the legs yellow.

WE ſhall now proceed to enumerate the ſpecies of fig-eaters that are found in America. They are in general larger than thoſe of the ancient continent. We have already noticed their diſtinguiſhing character, and we can only ſubjoin ſome details with regard to their habits. They are of a wandering diſpoſition; they paſs the ſummer in Carolina, or even ſo far north as Canada, and return to the warmer regions to breed and raiſe their young. They inhabit the cleared grounds and the cultivated ſpots: they perch on ſmall ſhrubs, and feed on inſects and ripe tender fruits, ſuch as thoſe of the bananas, of the mangroves, and of the fig-trees, which are not natives of that climate, but were tranſported thither; they enter the gardens to peck them, and hence their name: however, they on the whole eat more inſects than fruits; for, if theſe are hard, they cannot break them.

THE

SPOTTED FIG-EATER.

Le Figuier Tachete, *Buff.*

FIRST SPECIES.

Ficedula Canadensis, *Briss.*

THIS bird is seen in Canada during summer, but makes only a short stay, and does not breed there: its ordinary residence is in Guiana, and other parts of South America. Its warble is pleasant, and much like that of the linnet.

The head and all the upper side of its body are of a fine yellow, with reddish spots on the lower part of the neck, and on the breast and sides: the upper surface of its body, and the superior coverts of its wings, are of an olive green: the quills of its wings are brown, and edged exteriorly with the same green: the quills of the tail are brown, and bordered with yellow: the bill, the legs, and the nails are blackish.

A variety of this species, or perhaps the female, is represented in the same plate; for it differs from the other only because the upper

 side

ſide of the head is, like the body, of an olive green: but theſe differences are inſufficient to form a ſeparate ſpecies.

THE

RED-HEADED FIG-EATER.

LE FIGUIER A TETE ROUGE, *Buff.*

SECOND SPECIES.

Motacilla Petechia, *Linn. & Gmel.*
Sylvia Petechia, *Lath. Ind.*
Ficedula Erythrocephalus, *Briſſ.*
The Yellow Red-poll, *Edw.*
The Red-headed Warbler, *Penn. & Lath.*

THE crown of the head is of a beautiful red: all the upper ſide of the body is olive green; the under ſide of a fine yellow, with red ſpots on the breaſt and belly: the wings and tail are brown: the bill is black, and the legs are reddiſh. The female has no difference from the male, except that its colours are not ſo bright. It is a ſolitary, tranſient bird: it arrives in Pennſylvania in the month of March, but does not breed there: it frequents the

the brakes, ſeldom perches on large trees, and it feeds on the inſects which it finds on the ſhrubs. [A]

[A] Specific character of the *Motacilla Petechia* : " It is olive ; below yellow, with yellow red drops ; has a red cap."

THE

WHITE-THROATED FIG-EATER.

LE FIGUIER A GORGE BLANCHE, *Buff.*

THIRD SPECIES.

Motacilla Albicollis, *Gmel.*
Sylvia Albicollis, *Lath. Ind.*
Ficedula Dominicenſis, *Briſſ.*
The Saint Domingo Warbler, *Lath. Syn.*

THIS bird is found in St. Domingo. In the male, all the upper ſurface of the body, and the ſmall ſuperior coverts of the wings, are of an olive green : the ſides of the head and throat are whitiſh : the lower part of the neck and breaſt is yellowiſh, with ſmall red ſpots : the reſt of the upper ſide of the body is yellow : the great ſuperior coverts, and the quills of the wings,

wings, and thoſe of the tail, are brown, and edged with olive yellow: the bill, the legs, and nails are brown gray.

The female differs not from the male, except that the green on the upper part of the neck is mixed with cinereous.

THE

YELLOW-THROATED FIG-EATER.

LE FIGUIER A GORGE JAUNE, *Buff.*

FOURTH SPECIES.

Motacilla Ludoviciana, *Gmel.*
Sylvia Ludoviciana, *Lath. Ind.*
Ficedula Ludoviciana, *Briſſ.*
The Louiſiane Warbler, *Lath. Syn.*

THIS bird is a native of Louiſiana and of Saint Domingo. In the male, the head and all the upper ſide of the body are of a fine olive green, which is ſlightly tinged with yellowiſh on the back: the ſides of the head are of a dilute cinereous: the throat, the lower part of the neck, and the breaſt, are of a fine yellow,

yellow, with ſmall reddiſh ſpots on the breaſt: the reſt of the under ſide of the body is of a yellowiſh white: the ſuperior coverts of the wings are blueiſh, and terminated with white, which forms two croſs white bars on each: the quills of the wings are of a blackiſh brown, and edged exteriorly with blueiſh cinereous, and white within: the three firſt quills on each ſide have alſo a white ſpot on the extremity of their inſide: the upper mandible is brown; the lower gray; the legs and nails aſh-coloured.

The plumage of the female is the ſame with that of the male, only there are no red ſpots on the breaſt.

We cannot help obſerving that Briſſon has confounded this bird with the *pine-creeper* of Edwards, which is indeed a fig-eater, but different from the preſent. We ſhall notice it afterwards.

THE

GREEN AND WHITE FIG-EATER.

LE FIGUIER VERT & BLANC, *Buff.*

FIFTH SPECIES.

Motacilla Chloroleuca, *Gmel.*
Sylvia Chloroleuca, *Lath. Ind.*
Ficedula Dominicensis Minor, *Briss.*
The Green and White Warbler, *Lath. Syn.*

THIS is alſo a native of Saint Domingo. The head and the under ſide of the neck are of a yellowiſh aſh colour in the male; the ſmall ſuperior coverts of the wings, and all the upper ſide of the body, olive green; the throat and all the under ſide of the body yellowiſh white; the great ſuperior coverts of the wings and the quills brown, and edged with yellowiſh green; the quills of the tail of an exceeding deep olive green: the lateral ones have, on their inſide, a yellow ſpot, that is broader the more they are exterior: the bill, the legs, and nails are brown gray.

In the female the colours are fainter, which is the only difference.

THE

THE

ORANGE-THROATED FIG-EATER.

LE FIGUIER A GORGE ORANGEE, *Buff.*

SIXTH SPECIES.

Motacilla Auricollis, *Gmel.*
Sylvia Auricollis, *Lath. Ind.*
Ficedula Canadensis Major, *Briss.*
The Orange-throated Warbler, *Lath. Syn.*

BRISSON terms this the *Canada Fig-eater*; but, probably, like the rest of the genus, it is only a bird of passage in that climate. The head, the upper side of the neck, the back, and the small superior coverts of the wings, are of an olive green; the rump, and the great superior coverts of the wings, cinereous; the throat, the lower part of the neck, and the breast, orange; the belly, pale yellow; the lower belly, and the legs, whitish; the quills of the wings brown, and edged exteriorly with cinereous: the two middle quills of the tail are cinereous: all the rest are white within, and blackish on the outside, and at the tip.

There is no difference in the plumage between the male and the female, except that the colours in the latter are less vivid.

THE

THE

CINEREOUS-HEADED FIG-EATER.

LE FIGUIER A TETE CENDREE, *Buff.*

SEVENTH SPECIES.

Motacilla Maculosa, *Gmel.*
Sylvia Maculosa, *Lath. Ind.*
Ficedula Pennsylvanica Nævia, *Briss.*
The Yellow-rumped Warbler, *Lath. Syn.*

THIS bird was sent from Pennsylvania to England, and Edwards calls it the *Yellow-rumped Flycatcher*: he has very improperly given the appellation of *Flycatcher* to all the fig-eaters that he has described and delineated. In the present, the crown and sides of the head are cinereous: the upper surface of the neck and the back are of an olive green, spotted with black: the throat, the breast, and the rump, are of a fine yellow, with black spots on the breast: the superior coverts of the wings are of a deep ash colour, and terminated with white, which forms two transverse white bars on each wing: the quills of the wing are deep cinereous, edged with white: the two middle quills of the tail are black; the others are blackish, with

with a great white ſpot on the inſide: the bill, the legs, and the nails are brown.

THE

BROWN FIG-EATER.

LE FIGUIER BRUN, *Buff.*

EIGHTH SPECIES.

Motacilla Fuſcenſis, *Gmel.*
Sylvia Fuſcenſis, *Lath. Ind.*
Ficedula Jamaicenſis, *Briſſ.*
Muſcicapa Pallide-Fuſca, *Ray.*
Luſcinia Muſcicapa Pallidè-Fuſca, *Klein.*

SIR Hans Sloane is the firſt who mentions this bird, which he found in the cultivated parts of Jamaica, and which he calls *Worm-eater.* The head, the throat, all the upper ſide of the body, the wings and the tail, are light brown: the under ſide of the body is variegated with the ſame colours as the plumage of the larks. This is all that author ſays on the ſubject. [A]

[A] Specific character of the *Motacilla Fuſcenſis*: "It is duſkiſh; below variegated with blackiſh and rufous gray; the bill, the throat, and a bar at the eyes, brown."

THE

BLACK-CHEEKED FIG-EATER.

LE FIGUIER AUX JOUES NOIRES, *Buff.*

NINTH SPECIES.

Turdus-Trichas, *Linn.* & *Gmel.*
Sylvia-Trichas, *Lath. Ind.*
Ficedula Marilandica, *Briſſ.*
The Maryland Yellow-throat, *Edw.*
The Yellow-breaſted Warbler, *Penn.* & *Lath.*

WE are indebted to Edwards for the account of this bird. It inhabits Pennſylvania, and frequents the ſmall woods that are watered by rills, at the ſides of which it is commonly found. It only ſpends the ſummer in that climate, and diſappears before the winter; which ſhews that this fig-eater, like the others, is only a bird of paſſage in thoſe parts of North America.

The ſides of its head are of a fine black, and the crown is reddiſh brown: the upper ſide of the neck, the back, the rump, and the wings are of a deep olive green; the throat and breaſt of a fine yellow; the reſt of the under ſide of the body pale yellow: the bill and legs are brown.

THE

THE

YELLOW SPOTTED FIG-EATER.

LE FIGUIER TACHETE DE JAUNE, *Buff.*

TENTH SPECIES.

Motacilla Tigrina, *Gmel.*
Sylvia Tigrina, *Lath. Ind.*
Ficedula Canadensis Fusca, *Briss.*
The Spotted Yellow Flycatcher, *Edw. Penn. & Lath.*

WE borrow the description of this bird also from Edwards. Both the male and female were caught at sea eight or ten leagues off Saint Domingo, in the month of November, and brought to England by the same ship. The author observes properly that these are migratory birds, and were then on their passage from North America to the island of Saint Domingo.

The head and all the upper side of the body are olive: above the eyes there is a yellow bar; the throat, the lower part of the neck, the breast, and the inferior coverts of the wings, of a fine yellow, with little black spots: the belly and the legs are of a pale yellow, without spots; the wings and tail of a dull olive green: there is a long

 white

white ſpot on the ſuperior coverts of the wings; and the lateral quills of the tail are white one half of their length.

The female has no difference from the male, except that the breaſt is whitiſh, with brown ſpots; and that the olive green of the upper ſurface of the body is not ſo gloſſy. Briſſon has taken the female for another ſpecies, which he has termed *the brown* fig-eater of Saint Domingo.

THE

BROWN AND YELLOW FIG-EATER.

LE FIGUIER BRUN & JAUNE, *Buff.*

ELEVENTH SPECIES.

Motacilla Trochilus.
Motacilla Acredula, *Linn.*
Ficedula Carolinenſis, *Briſſ.*
Œnanthe Fuſco-lutea Minor, *Ray.*
The Yellow Titmouſe, *Cateſby.*
The Yellow Wren, *Edw.*
The Scotch Wren, *Penn. & Lath.*

THIS bird is found in Jamaica. Sloane and Brown have both deſcribed it, and Edwards has given a coloured figure under the name

name of *Yellow Wren*, which is improper. Catesby and Klein have fallen into another miſtake, reckoning it a titmouſe. It breeds in Carolina, but does not continue there during the winter: the head, all the upper ſurface of the body, the wings, and the tail, are of a greeniſh brown: there are two ſmall bars on each ſide of the head: all the under ſurface of the body is of a fine yellow: the ſuperior coverts of the wings are terminated with green and light olive, which forms two oblique bars in each: the quills of the wings are edged exteriorly with yellow; the bill and legs are black *.

* This bird, which ſeems to be only a variety of the yellow wren, breeds in North Carolina, and retires in winter to Jamaica. It occurs alſo in moſt parts of Europe, from India to Kamtſchatka.

THE

PINE FIG-EATER.

LE FIGUIER DES SAPINS, *Buff.*

TWELFTH SPECIES.

Certhia-Pinus, *Linn. & Gmel.*
Sylvia-Pinus, *Lath. Ind.*
Parus Americanus, *Briss.*
The Pine Warbler, *Penn. & Lath.*

EDWARDS calls this bird the *Pine-creeper*; but it does not belong to that genus, though it creeps on the pines in Carolina and Pennsylvania. The bill of the creepers, it is well known, is bent like a sickle; whereas it is straight in this bird, which resembles the fig-eaters so much in every other respect, that it ought to be classed with them. Catesby is also mistaken in ranging it with the titmice, probably because, like these, it creeps on trees. Brisson has committed an oversight in separating the *Pine-creeper* of Catesby from that of Edwards.

The head, the throat, and all the under side of the body, are of a very beautiful yellow: there is a small black bar on each side of the head.

head: the upper part of the neck, and all the upper surface of the body, are of a yellow green or shining olive, which is still more bright on the rump: the wings and tail are of a blueish iron colour: the superior coverts are terminated with white, which forms on each wing two transverse white bars: the bill is black, and the legs are of a yellowish brown.

The female is entirely brown.

This bird appears in Carolina in winter, where Catesby tells us that it searches on the deciduous trees for insects. It is also seen during summer in the northern provinces. Bartram informed Edwards, in a letter, that it arrives in Pennsylvania in the month of April, and continues all the summer: however, he confesses that he never saw its nest. It lives on the insects that lodge on the leaves, and in the buds of trees. [A]

[A] Specific character of the *Certhia-Pinus*: "It is yellow; above olive; its wings blue, with two white bars. It is four and a half inches long; its tail forked."

THE

BLACK-COLLARED FIG-EATER.

LE FIGUIER A CRAVATTE NOIRE, *Buff.*

THIRTEENTH SPECIES.

Motacilla Virens, *Gmel.*
Sylvia Virens, *Lath. Ind.*
Ficedula Pennſylvanica Gutture Nigro, *Briſſ.*
The Black-throated Green Flycatcher, *Edw.*
The Green Warbler, *Penn. & Lath.*

THIS fig-eater was ſent from Pennſylvania by Bartram to Edwards: it is a bird of paſſage in that climate, where it arrives in April, and advances northwards, and in September it returns again to the ſouth. It feeds on inſects, like all the others of this genus.

The crown of the head, all the upper ſurface of the body, and the ſmall ſuperior coverts of the wings, are of an olive green: the ſides of the head and neck are of a fine yellow; the throat, and under ſurface of the neck, black, which forms a ſort of collar of that colour: the breaſt is yellowiſh; the reſt of the under ſide of the body white, with ſome blackiſh ſpots on the flanks: the great ſuperior coverts of the wings are

are of a deep brown, and terminated with white, which forms on each wing two white tranſverſe bars: the quills of the wings, and thoſe of the tail, are of a deep aſh colour: the three outer ones on each ſide of the tail are marked with white ſpots within: the bill is black, and the legs brown.

THE

YELLOW-HEADED FIG-EATER.

LE FIGUIER A TETE JAUNE, *Buff.*

FOURTEENTH SPECIES.

Motacilla Icterocephala, *Gmel.*
Sylvia Icterocephala, *Lath. Ind.*
Ficedula Canadenſis Icterocephalos, *Briſſ.*
The Quebec Warbler, *Penn. & Lath.*

BRISSON is the firſt who deſcribed this bird. He tells us that it is found in Canada; but it is probable only migratory in that northern climate, like ſome other ſpecies of fig-eaters. The crown of its head is yellow: there is a great black ſpot on each ſide of the head,

 over

over the eyes, and another whitiſh one below them; the back of the head, the upper ſide of the neck, and all the upper ſide of the body, covered with black feathers, edged with yellowiſh green the throat, and all the under ſide of the body, whitiſh; the ſuperior coverts of the wings black, and tipped with yellowiſh, which forms on each wing two croſs yellowiſh bars: the quills of the wings and of the tail are blackiſh, and edged exteriorly with olive green and whitiſh; the inſide of the three lateral quills of the tail yellowiſh white, from their middle to their extremity: the bill, the legs, and nails, are blackiſh.

It appears that the bird, No. 731, Fig. 2. *Pl. Enl.* denominated the *Miſſiſſippi Fig-eater*, is only a variety of the preſent, occaſioned by age or ſex; for the only difference is, that it has no ſpots on the ſides of its head, and that its colours are not ſo deep. [A]

[A] Specific character of the *Motacilla Icterocephala*: "It is black; whitiſh below: has a yellow cap, and a black bar on its eyes, and two yellow ones on its wings."

THE

THE

YELLOW-THROATED CINEREOUS FIG-EATER.

LE FIGUIER CENDRE A GORGE JAUNE, *Buff.*

FIFTEENTH SPECIES.

Motacilla Dominica, *Linn. & Gmel.*
Sylvia Dominica, *Lath. Ind.*
The Jamaica Warbler, *Lath. Syn.*

DR. Sloane gives the account of this bird, which is found in Jamaica and Saint Domingo. The head, the whole of the upper ſurface of the body, and the ſmall ſuperior coverts of the wings, are of an aſh-colour: on each ſide of the head there is a yellow longitudinal bar; below the eyes a large black ſpot; on the outer edge of each eye a white ſpot: the throat, the under ſide of the neck, the breaſt, and belly, are yellow, with ſome ſmall black ſpots on each ſide of the breaſt: the great ſuperior coverts of the wings are brown, edged exteriorly with cinereous, and tipped with white, which forms two tranſverſe white bars on each wing: the quills of the wings and of the

the tail are of an aſh brown, and edged exteriorly with gray: the two outer quills on each ſide of the tail are marked with a white ſpot near the end of their interior ſurface: the bill, the legs, and the nails are brown. [A]

[A] Specific character of the *Motacilla Dominica*: "It is cinereous; below white; a yellow ſpot before the eyes; white behind, and black below."

THE

COLLARED CINEREOUS FIG-EATER.

LE FIGUIER CENDRE A COLLIER, *Buff.*

SIXTEENTH SPECIES.

Parus Americanus, *Linn. & Gmel.*
Ficedula Carolinenſis Cinerea, *Briſſ.*
The Finch-Creeper, *Cateſby.*
The Creeping Titmouſe, *Penn.*

WE are indebted to Cateſby for the account of this bird, which he terms the *Finch-creeper*; but it belongs to neither of theſe genera, and is really a fig-eater. It is found in North America, from Carolina to Canada.

The

The head, the upper ſide of the neck, the rump, and the ſuperior coverts of the wings, are of an aſh colour: the back is olive green; the throat and breaſt yellow, with a half collar of cinereous on the lower part of the neck: the reſt of the under ſide of the body is white, with ſome ſmall red ſpots on the flanks: the great ſuperior coverts of the wings are tipped with white, which forms on each wing two tranſverſe white bars: the quills of the wings and of the tail are blackiſh: the two exterior feathers, on each ſide of the tail, have a white ſpot at the termination of their inner ſurface: the upper mandible of the bill is brown; the lower mandible and the legs, yellowiſh.

Theſe birds creep on the trunks of large trees, and feed on the inſects which they pick out of the cracks in the bark: they continue during the whole winter in Carolina. [A]

[A] Specific character of the *Parus Americanus*: "It is blueiſh; its temples, its breaſt, and its back yellowiſh; its flanks purpliſh."

THE

BELTED FIG-EATR.

LE FIGUIER A CEINTURE, *Buff.*

SEVENTEENTH SPECIES.

Motacilla Canadenſis, *Linn.*
Motacilla Cincta, *Gmel.*
Sylvia Cincta, *Lath.*
Ficedula Canadenſis Cinerea, *Briſſ.*
The Belted Warbler, *Penn.* & *Lath.*

ON the crown of the head there is a yellow ſpot, and, on each ſide, a white bar: the reſt of the head, the upper ſurface of the body, the ſuperior coverts of the wings, are of a deep cinereous, almoſt black. But the moſt obvious character is a yellow belt between the breaſt and belly, which are both white, variegated with ſome ſmall brown ſpots: the great ſuperior coverts of the wings are tipped with white, which forms two white tranſverſe bars on each wing: the ſuperior coverts of the tail are yellow: the quills of the wings and of the tail are brown: the two exterior quills, on each ſide of the tail, have a white ſpot near the termination of of the inner ſurface: the bill is black; the legs and nails brown.

The female differs not from the male, except that the upper ſurface of the body is brown, and the ſuperior coverts of the tail are not yellow. [A]

[A] Specific character of the *Motacilla Canadenſis*: "Above blue; below white; its throat, and the quills of its wings and tail, black."

THE

BLUE FIG-EATER.

LE FIGUIER BLEU, *Buff.*

EIGHTEENTH SPECIES.

Motacilla Canadenſis, *Linn. & Gmel.*
Sylvia Canadenſis, *Lath.*
Ficedula Canadenſis Cinerea Major, *Briſſ.*
The Black-throated Warbler, *Penn. & Lath.*

THIS is the *blue fly-catcher* of Edwards. It was caught at ſea eight or ten leagues ſouth of Saint Domingo; but, from his ſtatement, he ſeems to have received another of the ſame

ſame birds from Pennſylvania. They arrive in that province about May, and remain through the ſummer: and thus they are migratory in North America, as are almoſt all the other fig-eaters, whoſe native climate is South America.

The head, all the upper part of the body, and the ſuperior coverts of the wings, are of a blue ſlate colour: the throat, and the ſides of the head and neck, are of a fine black; the reſt of the under ſurface of the body whitiſh; the quills of the wings, and of the tail, blackiſh, with a white ſpot on the great quills of the wings: the bill and legs are black: in the *Planches Enluminées* theſe are yellow: perhaps the ſmall ſcales of the dried ſpecimen from which it was deſigned, were worn off.

THE

THE

VARIEGATED FIG-EATER.

LE FIGUIER VARIE.

NINETEENTH SPECIES.

Motacilla Varia, *Linn.* & *Gmel.*
Sylvia Varia, *Lath.*
Ficedula Dominicensis Varia, *Briss.*
The Black and White Creeper, *Edw.*
The White-poll Warbler, *Penn.* & *Lath.*

SLOANE found this bird in Jamaica, and Edwards received a specimen from Pennsylvania, where it appears in the month of April. It feeds on insects, and continues during the summer; but, on the approach of winter, it returns to the tropical parts of the American continent. The crown of its head is white; the sides black, with two small white bars: the back and rump are white, variegated with large black spots: the throat is black; the breast and belly white, with some black spots on the breast and sides: the great superior coverts of the wings are black, tipped with white, which forms two transverse white bars on each wing: the quills of the wings are gray, and edged

edged with white on the inſide: the quills of the tail are black, and edged with iron gray: the lateral ones have white ſpots on their inſide: the bill and legs are black. [A]

[A] Specific character of the *Motacilla Varia*: "It is ſpotted with black and white; has two white ſtripes on the wings; its tail forked."

THE

RUFOUS-HEADED FIG-EATER.

LE FIGUIER A TETE ROUSSE, *Buff.*

TWENTIETH SPECIES.

Motacilla Ruficapilla, *Gmel.*
Sylvia Ruficapilla, *Lath.*
Ficedula Martinicana, *Briſſ.*
The Bloody-ſide Warbler, *Lath.*

THIS bird was ſent from Martinico to M. Aubry, rector of Saint Louis. The head is rufous; the upper part of the neck, and all the upper ſide of the body, olive green; the throat and breaſt yellow, variegated with longitudinal rufous ſpots; the reſt of the under ſide of the body, light yellow without ſpots: the ſuperior coverts and the quills of the wings and of the tail are brown, edged with olive green: the two exterior quills on each ſide of the tail are light yellow on the inſide; the bill brown, and the legs gray.

We conceive that the bird mentioned by Father Feuillée under the appellation of *chloris erithacorides*, is the ſame with this. Its bill is black, according to that author, and pointed with a minute portion of blue at the root of the lower mandible: its eye is of a fine gloſſy black, and the head and the nape of the neck are of a tawny colour, like dry leaves: all the fore ſide of the bird is yellow, ſtreaked, like the European thruſhes, with the colour of the head: all the back is greeniſh, but the wings are black, and its mantle is alſo black: the feathers of the wings have a green edging: the thighs and the upper part of the legs are gray; but the under part is entirely white, mixed with a little yellow; and the toes are furniſhed with ſmall black nails, that are very ſharp *.

This bird is continually on the wing, and never reſts unleſs to feed: its ſong is ſlender, but mellow.

* *Obſervations Phyſiques du P. Feuillee,* p. 113.

THE

RED-BREASTED FIG-EATER.

Le Figuier a Poitrine Jaune, *Buff.*

TWENTY-FIRST SPECIES.

Motacilla Pennſylvanica, *Linn.* & *Gmel.*
Sylvia Pennſylvanica, *Lath. Ind.*
Ficedula Pennſylvanica Icterocephalos, *Briſſ.*
The Red-throated Flycatcher, *Edw.*
The Bloody-ſide Warbler, *Penn.*
The Red-throated Warbler, *Lath.*

EDWARDS has given the cock and hen of this bird, which he received from Pennſylvania, where they only pay a tranſient viſit in the ſpring, in their way to ſpend the ſummer in the more northern provinces. They live on inſects and ſpiders.

The crown of the head is yellow, with white on each ſide, and a ſmall black bar below the eyes: the upper ſurface of the neck, and the ſuperior coverts of the wings, are blackiſh: the feathers on the upper part of the body, and the quills of the wings, are black, and edged with olive green: the top of the breaſt, and the ſides of the body, are of a deep red;

red; the throat and the belly whitiſh: the great ſuperior coverts of the wings are tipped with white, which forms two white tranſverſe bars on each wing: the bill and legs are black.

The female is diſtinguiſhed from the male, by having no black on the back of the head, nor red on the breaſt. [A]

[A] Specific character of the *Motacilla Pennſylvanica*: "It has a yellowiſh cap, and its flanks are blood-coloured."

THE

CÆRULEAN FIG-EATER.

LE FIGUIER GRIS DE FER, *Buff.*

TWENTY-SECOND SPECIES.

Motacilla Cærulea, *Linn. & Gmel.*
Sylvia Cærulea, *Lath. Ind.*
Ficedula Pennſylvanica Cinerea, *Briſſ.*
The Little Blue-gray Flycatcher, *Edw.*
The Cærulean Warbler, *Penn. & Lath.*

WE are indebted to Edwards alſo for the deſcription and hiſtory of this bird. He has given figures of the cock and hen, with the

 neſt.

neſt. They are found in Pennſylvania, where they arrive in the month of March, and remain through the ſummer, and again return to the ſouth.

The head, and all the upper part of the body, are iron gray: there is a black bar on each ſide of the head, above the eyes: all the under ſurface of the body is white: the wings are brown: the two outer quills on each ſide of the tail are white: the third, on each ſide, has a white ſpot near the extremity; and what remains of it, as well as all the other quills of the tail, is of the ſame colour with the upper part of the body.

The female has not the black bars on the ſides of the head, which is the only difference between it and the male.

Theſe birds begin in April to conſtruct their neſts with the ſhort woolly ſubſtance that ſurrounds the buds of trees, and with the down of plants: the outſide is compoſed of a flat grayiſh moſs or lichen, which they gather on the rocks: and there is an intermediate layer of horſe hair. The form is nearly that of a ſhort cylinder; cloſe below, and having its aperture above.

It would appear that the bird, No. 704, fig. 1. *Pl. Enl* and denominated *the black-headed fig-eater of Cayenne*, belongs to this ſpecies; for it is preciſely like the male deſcribed by Edwards, except that the head, the quills of the wings, and thoſe of the middle of the tail, are of a fine black; and this difference is only ſufficient to conſtitute two varieties. [A]

[A] Specific character of the *Motacilla Cærulea*: "Above it is blue; below white; its wings and tail black."

THE

GOLDEN-WINGED FIG-EATER.

LE FIGUIER AUX AILES DORÉES, *Buff.*

TWENTY-THIRD SPECIES.

Motacilla Chryſoptera, *Linn. & Gmel.*
Sylvia Chryſoptera, *Lath.*
Ficedula Pennſylvanica Cinerea Gutture Nigro, *Briſſ.*
The Golden-winged Flycatcher, *Edw.*
The Gold-winged Warbler, *Penn. & Lath.*

WE borrow the account of this bird too from Edwards. It arrives in Pennſylvania in the month of April, and halts only a few days; it penetrates farther north, and returns to paſs the winter in the milder climates.

The head is of a fine yellow, and there is a large ſpot of gold colour on the ſuperior coverts of the wings: the ſides of the head are white,

 with

with a broad black bar that ſurrounds the eyes: all the upper ſurface of the body, the wings, and the tail, are of a deep aſh-colour: the throat and the lower part of the neck are black: the reſt of the under ſurface of the body is white: the bill and legs are black. [A]

[A] Specific character of the *Motacilla Chryſoptera*: "It is dark cinereous; below white; its cap, and the ſpot on its wings, yellow; its throat black."

THE

GOLDEN-CROWNED FIG-EATER.

LE FIGUIER COURONNE D'OR, *Buff.*

TWENTY-FOURTH SPECIES.

The Golden-crowned Flycatcher, *Edw. & Lath.*

WE adopt the epithet *golden-crowned*, given by Edwards. It is a bird of paſſage in Pennſylvania, where it appears in the ſpring; and, after halting a few days, it advances farther north; and, on the approach of winter, it returns to the warmer climates.

On the crown of the head there is a round ſpot of a fine gold colour: the ſides of the head, the

the wings, and the tail, are black: the upper part of the neck, the back, and the breaſt, are of a blue ſlate colour, ſpotted with black, and the ſides of the body are yellow, with ſome black ſpots: all the under ſurface of the body is whitiſh: the great ſuperior coverts of the wings are tipped with white, which forms two tranſverſe white bars on each wing: the bill and legs are blackiſh.

The female differs not from the male, except that the upper ſurface of the body is brown, and there is no black on the ſides of the head, nor on the breaſt.

THE

ORANGE FIG-EATER.

LE FIGUIER ORANGÉ.

TWENTY-FIFTH SPECIES.

Motacilla Chryſocephala, *Gmel.*
Sylvia Chryſocephala, *Lath. Ind.*
The Orange-headed Warbler, *Lath. Syn.*

THIS ſpecies is new: it is found in Guiana, whence it was ſent to the king's cabinet. The crown and ſides of its head, the throat, the ſides, and under ſurface of its neck, are of a beautiful orange colour, with two ſmall brown bars on each ſide of the head. All the upper ſurface of the body, and the quills of the wings, are of a reddiſh brown: the ſuperior coverts of the wings are variegated with black and white: the breaſt is yellowiſh as well as the belly: the quills of the tail are black, and edged with yellowiſh: the bill is black, and the feet are yellow.

THE

CRESTED FIG-EATER.

LE FIGUIER HUPPE, *Buff.*

TWENTY-SIXTH SPECIES.

Motacilla Criſtata, *Gmel.*
Sylvia Criſtata, *Lath. Ind.*
The Creſted Warbler, *Lath. Syn.*

THIS bird has not been noticed by any naturaliſt. It is found in Guiana, where it is probably ſtationary, ſince it is ſeen at all ſeaſons: it lives in cleared parts, feeds on inſects, and has the ſame habits and œconomy with

with the other fig-eaters. The under ſide of its body is gray, mixed with whitiſh: and the upper is brown, ſhaded with green: it is diſtinguiſhed from the other fig-eaters by its creſt, which conſiſts of ſmall round feathers, half erect, fringed with white, on a blackiſh brown ground, and briſtled as far as the eye and the root of the bill. It is four inches long, including the tail: the bill and legs are of a yellowiſh brown. [A]

[A] Specific character of the *Motacilla Criſtata*: "Above it is duſky green; below greeniſh gray; the creſt on its head blackiſh brown, white at the margin."

THE

BLACK FIG-EATER.

Le Figuier Noir, *Buff.*

TWENTY-SEVENTH SPECIES.

Motacilla Multicolor, *Gmel.*
Sylvia Multicolor, *Lath. Ind.*
The Rufous and Black Warbler, *Lath. Syn.*

ANOTHER ſpecies, which is alſo found at Cayenne, but which is rarer, is the black fig-eater; ſo termed, becauſe the head and

and throat are ſhrouded with black, which extends over the top and ſides of the neck, and on the wings, and backwards as far as the origin of the tail: the ſame black appears again in a broad bar on the tips of the quills, of which the firſt half is of a bay colour: there is a ſhort ſtreak of the ſame colour on the ſix or ſeven firſt quills of the wing, near their inſertion, and on the ſides of the neck and breaſt: the fore part of the body is whitiſh gray: the bill and legs are yellowiſh brown. This is one of the largeſt of the fig-eaters, for it is near five inches long.

THE

OLIVE FIG-EATER.

LE FIGUIER OLIVE, *Buff.*

TWENTY-EIGHTH SPECIES.

Motacilla Æquinoctialis, *Gmel.*
Sylvia Æquinoctialis, *Lath. Ind.*
The Æquinoctial Warbler, *Lath. Syn.*

THIS is another fig-eater which is pretty common in Cayenne, where it is ſtationary. All the upper part of the body, and of the head, is of an olive green on a brown ground; the ſame olive ſtrikes alſo through the blackiſh brown of the wings and tail: the lower part of the throat and breaſt, as far as the belly, is of a light yellow. It is alſo one of the largeſt of the fig-eaters, for it is near five inches long.

THE

PROTHONOTARY FIG-EATER.

LE FIGUIER PROTHONOTAIRE, *Buff.*

TWENTY-NINTH SPECIES.

Motacilla-Protonotarius, *Gmel.*
Sylvia-Protonotarius, *Lath.*
The Prothonotary Warbler, *Penn. & Lath.*

THIS bird is called *Prothonotary* in Louiſiana, and we ſhall retain the name, in order to diſcriminate it from the other fig-eaters. The head, the throat, the neck, the breaſt, and the belly, are of a fine jonquil yellow; the back olive; the rump cinereous; the inferior coverts of the tail white; the quills of the

the wings, and of the tail, blackiſh and cinereous; the bill and the legs black.

Beſides the foregoing twenty-nine ſpecies of fig-eaters, which are all natives of the New World, there are five ſpecies or varieties in Louiſiana alone. The ſpecimens are preſerved in M. Mauduit's cabinet, and were brought by Le Beau, king's phyſician in Louiſiana.

THE

HALF-COLLARED FIG-EATER.

LE FIGUIER A DEMI-COLLIER, *Buff.*

THIRTIETH SPECIES.

Motacilla Semitorquata, *Gmel.*
Sylvia Semitorquata, *Lath. Ind.*
The Half-collared Warbler, *Penn. & Lath.*

THIS little bird is of a very light aſh colour below the throat, and on all the under ſurface of the body: there is a yellowiſh half

half collar on the lower part of the neck: the upper ſide of the head is olive, bordering on yellow: there is a cinereous bar behind the eyes: the ſuperior coverts of the wings are brown, edged with whitiſh; and the middle quills are alſo brown, but edged with olive, and tipped with white: the belly has a yellowiſh tint: the quills of the tail are cinereous: the two middle ones have no white: the four on each ſide of theſe are edged with white on the inſide: all the ten are pointed at the end: the bill is blackiſh above, and whitiſh below. The bird is four inches and a half long; the tail twenty-one lines, and projecting ten lines beyond the wings: the legs are blackiſh.

THE

YELLOW-THROATED FIG-EATER.

LE FIGUIER A GORGE JAUNE, *Buff.*

THIRTY-FIRST SPECIES.

Motacilla Fulva, *Gmel.*
Sylvia Fulva, *Lath. Ind.*
The Orange-bellied Warbler, *Penn. & Lath.*

THE throat, the neck, and the top of the breaſt, are yellow; only the top of the breaſt is a little more duſky: the reſt of the under ſide of the body is ruſty, verging to yellow on the inferior coverts of the tail: the head and the upper ſide of the body are brown olive: the ſmall inferior coverts of the wings are yellow, variegated with brown, which forms a diſtinct yellow border: the quills of the wings are brown; the middle ones edged with olive, and the great ones with light gray, which, growing more dilute, becomes white on the firſt quill: thoſe of the tail are brown, edged with olive: the bill is brown above, and lighter brown below: the legs are of a yellowiſh brown.

THE

THE

OLIVE-BROWN FIG-EATER.

Le Figuier Brun-Olive, *Buff.*

THIRTY-SECOND SPECIES.

Motacilla Fuſca, *Gmel.*
Sylvia Fuſca, *Lath. Ind.*
The Olive-brown Warbler, *Penn. & Lath.*

THE upper ſide of the head, of the neck, and of the body, is brown, verging on olive; the ſuperior coverts of the tail olive: the throat, the fore part of the neck, the breaſt, and the flanks, are whitiſh, and variegated with gray ſtreaks: the belly is yellowiſh white: the inferior coverts of the tail are entirely yellow: the ſuperior coverts of the wings, and their middle quills, are brown, edged with a lighter brown, and tipped with whitiſh: the great quills of the wings are brown, edged with light gray: the quills of the tail are alſo brown, edged with light gray, and with a yellowiſh tint on the middle ones: the two lateral ones on each ſide are marked with a white ſpot at the extremity of their interior

terior ſurface, and the firſt on each ſide is tipped with white: the bill is brown above, and of a diluter brown below: the legs are brown.

THE

GRASSET FIG-EATER.

LE FIGUIER GRASSET, *Buff.*

THIRTY-THIRD SPECIES.

Motacilla Pinguis, *Gmel.*
Sylvia Pinguis, *Lath.*
The Graſſet Warbler, *Penn.* & *Lath.*

THE upper ſide of the head and of the body is of a deep greeniſh gray, or of a coarſe olive green, with a yellow ſpot on the head, and black ſtreaks on the body: the rump is yellow: the throat, and the under ſide of the neck, are of a ruſty colour, through which the deep cinereous ground appears: the reſt of the under ſide of the body is whitiſh: the great quills of the wings are brown, edged exteriorly with gray, and interiorly with whitiſh: the middle quills are blackiſh, edged exteriorly and tipped

tipped with gray: the quills of the tail are black, edged with gray: the four lateral quills are each marked with a white ſpot near the end of their interior ſurface: the bill and legs are black.

THE

ASH-THROATED CINEREOUS FIG-EATER.

Le Figuier Cendre a Gorge Cendree, *Buff.*

THIRTY-FOURTH SPECIES.

Motacilla Cana, *Gmel.*
Sylvia Cana, *Lath. Ind.*
The Gray-throated Warbler, *Penn. & Lath.*

THE head and the upper ſide of the body are cinereous: the throat and all the under ſide of the body are of a lighter cinereous: the quills of the wings are cinereous, edged with whitiſh; the quills of the tail black; but the firſt on each ſide is almoſt entirely white: the ſecond quill is white on the half next the end:

the third is only tipped with white : the bill is black above, and gray below.

This and the preceding are called *graſſets* in Louiſiana, becauſe they are very fat *(gras)*. They perch on the tulip-trees, particularly on the magnolia, which is an evergreen ſpecies. [A]

[A] Specific character of the *Motacilla Cana* : "It is cinereous ; its wing quills whitiſh ; its tail quills black ; the outermoſt entirely white."

THE

GREAT FIG-EATER OF JAMAICA.

LE GRAND FIGUIER DE LA JAMAIQUE, *Buff.*

THIRTY-FIFTH SPECIES.

Motacilla-Calidris, *Linn. & Gmel.*
Sylvia-Calidris, *Lath. Ind.*
Ficedula Jamaicenſis Major, *Briſſ.*
The Hang-neſt Warbler, *Lath. Syn.*

EDWARDS was the firſt who deſcribed this bird. He terms it the *American Nightingale.* But it is by no means a nightingale, and it has all the characters of the fig eaters, with which Briſſon has properly ranged it. The upper

upper mandible is blackiſh; the lower fleſh-coloured: the upper ſurface of the back, of the head, and of the wings, is brown, with an obſcure tinge of greeniſh: the edges of the quills are of a lighter greeniſh yellow: an orange colour predominates on the under ſide of the body, from the throat to the tail: the inferior coverts of the wings, and all thoſe of the tail, and alſo the inner webs of the quills, are of the ſame colour. From the angle of the bill a black ſtreak ſtretches acroſs the eye; another extends below it: between theſe two, and under them, the orange forms two bars: the legs and toes are blackiſh. The bird is nearly as large as the red-breaſt, and not quite ſo thick. Edwards remarks that it bears great reſemblance to what Sloane, in his Natural Hiſtory of Jamaica, calls the *Icterus Minor, nidum ſuſpendens*. [A]

[A] Specific character of the *Motacilla-Calidris*: "Above it is greeniſh brown; below fulvous; a line upon and under the eyes black." It is of the ſize of the red-breaſt.

We cannot omit noticing three birds which our nomenclators have confounded with fig-eaters, but which are undoubtedly of a different kind.

 Theſe

Theſe are, 1. *The Great Fig-eater of Jamaica*, mentioned by Briſſon in his Supplement: its bill is entirely different from that of the fig-eaters.

2. The *Pennſylvanian Fig-eater* *, which is alſo diſtinguiſhed from the fig-eaters by its bill, and appears to be of the ſame genus with the preceding.

3. *The Great Fig-eater of Madagaſcar*, in the ornithology of the ſame author, which has rather the bill of the blackbird than of the fig-eater.

* The *Motacilla Vermivora* of Gmelin, or the *Worm-eater*, which is the ſubject of the next article.

THE

THE

MIDDLE-BILLS.

Les Demi-Fins, *Buff.*

WHEN we compare the birds which inhabit the two continents, we perceive that thoſe with ſtrong bills, and which feed upon grain, are the moſt numerous in the old; but, on the contrary, thoſe which have ſlender bills, and ſubſiſt upon inſects, predominate in the new. This circumſtance ſtrikingly evinces the extenſive influence which human induſtry has upon the productions of nature: for it is obviouſly the cultivation of the various ſorts of corn, ſubſervient to the ſupport of man, that has multiplied the granivorous birds. In the vaſt deſerts of America, in her magnificent foreſts, and her immenſe ſavannas, where rude unaſſiſted nature yields nothing ſimilar to our corn, and affords only fruits and ſmall ſeeds, with enormous quantities of inſects, there the number of the ſpecies of birds which feed on theſe, and have ſlender bills, is proportionally great: but an inſenſible gradation connects the various productions of the univerſe, and bids defiance to the ſhackles of ſyſtem.

Between the birds with ſtrong bills and thoſe with ſlender ones there is an intermediate claſs, which, though it has never been admitted by nomenclators*, has a real exiſtence. It comprehends thoſe birds in the New World which have ſtronger bills than the *pipits*, but not ſo ſtrong as the tanagres; and alſo thoſe birds of the old continent which have ſtronger bills than the *fauvettes*, but not ſo ſtrong as the larks. We might refer to it not only the calandre and other larks, but many ſpecies which have been ranged in other claſſes, becauſe this was not yet formed. Laſtly, the titmice will occupy the place between the middle-bills and thoſe which have ſlender bills: for though they appear delicate, yet, if we compare their thickneſs with their ſhortneſs, and conſider that they can break a nut, and pierce the head of much larger birds, we ſhall be convinced that they have conſiderable ſtrength.

* When this article was ſent to the preſs, I found that Edwards, in his Catalogue of Birds, &c. at the end of his ſeventh volume, has reckoned, among thoſe with bills of middle thickneſs, the following birds:

1. His Scarlet Bird.
2. His Red Summer Bird.
3. His White-faced Manakin.
4. His American Hedge-ſparrow.
5. His Indian Red-tail.
6. His Olive Flyca cher.
7. His Worm-eater.

THE

THE

WORM-EATER.

Le Demi-Fin, Mangeur de Vers, *Buff.*
Motacilla Vermivora, *Gmel.*
Sylvia Vermivora, *Lath. Ind.*
Ficedula Pennſylvanica, *Briſſ.*

THIS bird is entirely different from another worm-eater mentioned by Sloane: it is diſcriminated by its climate, and by its natural qualities. The bill is pretty ſharp, brown above, and fleſh-coloured below: its head is orange, and on each ſide there are two black bars, of which the one paſſes acroſs the eye, and the other below it; and they are ſeparated by a yellowiſh bar, beyond which they join near the back of the head: the throat and the breaſt are alſo of an orange colour, but which grows more dilute, as it diverges from the anterior parts, and is only whitiſh on the inferior coverts of the tail: the upper ſide of the neck, the back, the wings, and the tail, are of a deep olive green: the inferior coverts of the wings are of a yellowiſh white: the legs are fleſh-coloured.

This bird is found in Pennſylvania, where it

is migratory, as are all thoſe with ſlender bills, and ſome of thoſe which have ſtrong bills. It arrives in the month of July, and advances towards the north; but it is not ſeen, in that province, to return again in autumn; and the ſame is the caſe with all the others which paſs in the ſpring. Edwards ſuppoſes that their route to the ſouth lies beyond the Apalachian mountains; and undoubtedly they muſt be induced to change their track, by the greater abundance of inſects and worms which the back countries then afford.

This worm-eater is ſomewhat larger than the black-cap. [A]

[A] Specific character of the *Motacilla Vermivora*: "It is olive; its head, throat, and breaſt fulvous; the ſtripe on its eyes, and the arch of its eyebrows, black; a yellowiſh line above the eyes; the vent cinereous."

THE

BLACK AND BLUE MIDDLE-BILL.

Le Demi-Fin Noir & Bleu, *Buff.*
Fringilla Cyanomelas, *Gmel.*
The Blue-headed Finch, *Lath.*

KOELREUTER *, who firſt deſcribed this bird, ſays that it is a very rare ſpecies, and brought from India. He tells us that the bill is longer and more ſlender than in the finches †, and conſequently it ought to be referred to the claſs of the middle-bills.

Except the bill, which is brown, and the legs, which are alſo brown, but more dilute, this bird has only black and blue on its plumage: the black is ſpread on the throat, the bottom of the wing, and the fore part of the back, where it forms a ſemi-circle, whoſe convexity is turned towards the tail: beſides this, there is a black ſtreak which joins each noſtril to the eye on the ſame ſide: the quills of the wings are

* "Blue finch; its chin, its throat, the baſe of its wings, and the fore part of its back, black." *J. T. Koelreuter.* Peterſburg Tranſactions for 1765, p. 434.

† It is odd that, after making that aſſertion, he ſhould reckon it a finch.

blackiſh,

blackiſh, edged with blue, and this edging is broader in the middle ones: all the reſt of the plumage is varying blue, with copper-coloured reflexions.

This bird is nearly as large as the greater red-poll: its bill is five lines and a half long, and its tail conſiſts of twelve equal quills.

THE

THE

BLACK AND RUFOUS MIDDLE-BILL.

Le Demi-Fin Noir & Roux, *Buff.*
Motacilla Bonarienſis, *Gmel.*
Sylvia Bonarienſis, *Lath. Ind.*
The White-chinned Warbler, *Lath. Syn.*

COMMERSON* ſaw this bird at Buenos-Ayres. All the upper ſide of the head and body, from the baſe of the bill to the end of the tail, is of a diſtinct black: the throat, the fore part of the neck, and the flanks, are of a ruſt colour: there is ſome white between the forehead and the eyes, at the riſe of the throat, on the middle of the belly, at the bottom of the wings, and at the extremity of the exterior quills of the tail: the bill is blackiſh; the noſtrils very near its baſe, and half covered with ſmall feathers: the iris is cheſnut: the pupil is blackiſh blue; the tongue triangular, and not divided at the tip; laſtly, the hind nail is the ſtrongeſt of all.

* "Finch; black above from the front to the end of the tail; the throat, the lower part of the neck, and the belly, ferruginous; the middle of the lower belly, and the begining of the throat, whitiſh." *Commerſon.*

Commerſon,

Commerſon, induced no doubt by the ſhape of its bill, aſſigns this bird a place between the finches and the birds with ſlender bills *: and, for this reaſon, I have ranged it with the middle bills. The name of finch, according to Commerſon himſelf, does not correſpond with it, though he is obliged, for want of another term, to apply that. The bird is nearly as large as the linnet.

Total length five inches and two thirds; the bill five lines; the tail twenty-ſix lines; and conſiſts of twelve quills; and it projects twenty lines beyond the wings, which contain ſixteen or ſeventeen quills.

* M. Commerſon ſeems often to lean to the ſyſtem of Linnæus.

THE

THE

BIMBELE,

OR BASTARD LINNET.

Le Bimbelé, ou La Fauſſe Linotte, *Buff.*
Motacilla Palmarum, *Gmel.*
Sylvia Palmarum, *Lath. Ind.*
The Palm Warbler, *Lath. Syn.*

I OWE the information which I have obtained in regard to this bird to the Chevalier Le Fevre Deſhayes, who ſent me a drawing of it. It is called *bimbelé* by the negroes, from its reſemblance to an African bird of that name. But probably this appellation is not better applied than that of baſtard linnet; for the bird reſembles the linnet neither in its ſong, in its plumage, nor the ſhape of its bill.

Its ſong is not varied or rich: it turns on four or five notes. However, it is pleaſant: for the tones are full, ſoft, and mellow.

It lives on fruits and ſmall ſeeds. It prefers the haunt of the palms, and builds its neſt in a ſort

ſort of rooſt, which the palm birds and others form on thoſe trees, at the place where the foot-ſtalk ſupporting the cluſter is inſerted. It lays only two or three eggs, and this is perhaps one of the reaſons the *bimbelés* are ſo rare.

Its plumage is ſtill inferior to its ſong: the throat, the fore part of the neck, the breaſt, and the top of the belly, are dirty white, tinged with yellow: the legs, the lower belly, and the inferior coverts of the tail, are of a faint yellow; the flanks deep gray: all the upper part brown, deeper on the head, and lighter on the back: the rump and the ſuperior coverts of the tail are olive green: the quills, and the ſuperior coverts of the wings, and the quills of the tail, brown, edged exteriorly with a lighter colour: the two outer pairs of the tail quills edged interiorly with a broad bar of pure white near their extremity: the lower face of all theſe quills is of a ſlate gray: the iris is light brown.

The *bimbelé* weighs rather leſs than two gros and an half.

Total length five inches; the bill ſeven lines, very acute; the noſtrils oblong, with a protuberance; the alar extent ſeven inches; eighteen quills in each wing; the tail about eighteen

8 lines,

lines, compoſed of twelve quills nearly equal, and projects an inch beyond the wings. [A]

[A] Specific character of the *Motacilla Palmarum*: "It is brown; below dirty ochry white; the belly yellowiſh; the rump olive; the two outermoſt tail quills marked with a white ſtripe on the inſide near the tip."

THE

BANANA WARBLER.

Le Bananiſte, *Buff.*
Motacilla Bonanivora, *Gmel.*
Sylvia Bonanivora, *Lath.*

WE have already noticed a Jamaica bird among the finches, and termed it *bonana*, but which muſt be diſcriminated from this. The preſent is much ſmaller, its plumage is different, and, though it haunts the ſame tree, its habits are probably diſtinct. We might decide this matter, if the bird mentioned by Sloane were as well known as that which we are to deſcribe from a coloured drawing, and an account of it ſent by the Chevalier Deſhayes. It is found in Saint Domingo, and the negroes affirm that it ſuſpends its neſt: it is often ſeen on the bananas; but other birds alſo feed on the fruit of theſe trees, and therefore the name is not appropriated to it.

The banana warbler has a bill ſomewhat curved, very acute, and of a middle ſize. Beſides bananas, it feeds on oranges, citronelles, avigato pears, and papaws. We cannot decide whether it alſo eats ſeeds or inſects; but certainly

tainly there were no traces of theſe in the ſtomach of the one diſſected. It lodges in the banana plantations, in the grounds uncultivated and covered with buſhes. It flies by ſtarts and jerks, and its motion is rapid, and attended with a little noiſe. Its warble is ſcarcely varied: it is a ſeries of cadences that reſt more or leſs on the ſame tone.

But though the Bonana flies ſwiftly, Deſhayes regards it as too weak and delicate to perform diſtant journeys, and to ſupport the cold of the northern climates; he therefore concludes it to be a native of the new continent.

The upper ſide of the body is of a deep gray, almoſt blackiſh, which approaches to brown on the tail, and the coverts of the wings: the quills of the tail are not ſo deep coloured as thoſe of the wings, and are tipped with white: it has a ſort of white eyebrows: the eyes are placed in a black bar, which riſes from the bill, and melts into the dark colour of the back of the head: the throat is aſh gray: the breaſt, the belly, and the rump, are of a delicate yellow: the flanks, the thighs, and the inferior coverts of the tail, are variegated with light yellow and gray: ſome of the inferior coverts are white, and riſe on the tail: the anterior part of the ſhoulders is of a fine yellow: the bill is black; the legs ſlate gray.

 Total

Total length, three inches eight lines; the bill four lines; the noſtrils broad, and like an inverted creſcent, with a protuberance of the ſame ſhape, but of an oppoſite poſition: the tongue pointed; the *tarſus* ſeven lines; the alar extent ſix inches; the wings compoſed of ſeventeen quills; the tail fourteen or fifteen lines, and exceeds the wings about ſix or ſeven lines.

THE

THE

MIDDLE-BILL,

WITH WHITE CREST AND THROAT.

Le Demi-Fin à Huppe & Gorge Blanches, *Buff.*
Pipra Albifrons, *Linn. & Gmel.*
The White-faced Manakin, *Edw. & Lath.*

ALL that Edwards, who firſt deſcribed and figured this bird, mentions in regard to its hiſtory, is, that it is a native of South America and of the adjacent iſlands, ſuch as Cayenne. Its creſt conſiſts of white feathers, which are long, narrow, and pointed: in the placid ſtate they are reclined on the head, but when the bird is agitated by paſſion they become erect: the throat is white, edged with a black belt, which ſtretches from the one eye to the other: the back of the head, the fore part of the neck, the breaſt, the belly, the rump, the quills of the tail, their coverts both ſuperior and inferior, and the inferior coverts of the wings, are orange, which is more or leſs bright: the top of the back, the lower part of the neck joining the quills of the wings, their ſuperior coverts, and the thighs, are of a deep cinereous, verging more or leſs on

 blue:

blue: the bill is black, ſtraight, pretty acute, and of a middle ſize: the legs are orange yellow.

Total length, five inches and a quarter; the bill eight or nine lines; the *tarſus* ten lines; the outer toe connected almoſt its whole length to the mid toe; the tail compoſed of twelve quills, and projects eight or nine lines beyond the wings. [A]

[A] Specific character of the *Pipra Albifrons*: "It has a white creſt, its body brick coloured, its back black."

THE

THE

SIMPLE WARBLER, *Lath.*

L'Habit-Uni, *Buff.*
Motacilla Campeſtris, *Linn. & Gmel.*
Sylvia Campeſtris, *Lath. Ind.*
Curruca Sepiaria Jamaicenſis, *Briſſ.*
The American Hedge Sparrow, *Edw.*

EDWARDS regrets in ſome meaſure that the plumage of this bird is too ſimple and uniform, and that it has no peculiarity to characteriſe it. I ſhall adopt this very ſimplicity as the character. A ſort of cinereous cowl, with a ſlight tinge of green, covers the head and neck: all the upper ſide of the body, including the wings and the tail, is of a ruſty brown: the quills are aſh coloured beneath; the bill black, and the legs brown.

This bird is of the ſize of the hedge-ſparrow, but it is not of the ſame ſpecies, though Edwards has applied the ſame name; he expreſsly ſays that its bill is thicker and ſtronger. It is found in Jamaica. [A]

[A] Specific character of the *Motacilla Campeſtris*: "It is brown; its head greeniſh cinereous; its tail quills of the ſame colour with the body; its belly whitiſh."

THE

PITPITS.

THOUGH these birds bear great resemblance to the fig-eaters, and also inhabit the new continent, the difference is still so considerable, that they ought to be regarded as forming a distinct and separate genus. Most of the fig-eaters are migratory: all the pitpits continue settled in the hottest parts of America. They remain in the woods, and perch on the large trees; whereas the fig-eaters haunt only the cleared grounds, and lodge among the bushes, and on the middle-sized trees. The pitpits are also more social than the fig-eaters. they keep in large flocks, and mix familiarly with the small birds of other species: they are more joyous and lively, and are continually hopping. But besides the difference of their habits, they are also discriminated by their conformation: their bill is thicker, and not so slender as that of the fig-eaters; and for this reason we have placed the middle bills between them and the fig-eaters. The tail of the pitpits is also square-terminated, while, in the fig-eaters, it is somewhat forked. These two characters, drawn from

from the bill and the tail, are a ſufficient foundation for forming two genera of theſe birds.

We know only five ſpecies of the pitpits, and they are all found in Guiana and Brazil, and are nearly of the ſame ſize.

THE

GREEN PITPIT.

FIRST SPECIES.

Motacilla Cyanocephala, *Gmel.*
Sylvia Cyanocephala, *Lath. Ind.*
Sylvia Viridis, *Briſſ.*
The Blue-headed Warbler, *Lath. Syn.*

THE pitpits are in general nearly as large as the fig-eaters, but rather thicker: they are four and a half or five inches long. In the kind which we call the *green pitpit* the head and the ſmall ſuperior coverts of the wings only are of a fine blue, and the throat of a blueiſh gray; but all the reſt of the body and the great ſuperior coverts of the wings are of a brilliant green: the quills of the wings are brown, edged exteriorly with green: thoſe of the tail are of a duller green: the bill is brown, and the legs gray. It is pretty common at Cayenne.

THE

BLUE PITPIT.

SECOND SPECIES.

Motacilla Cayana, *Linn. & Gmel.*
Sylvia Cayana, *Lath. Ind.*
Sylvia Cayanenſis Cærulea, *Briſſ.*
The Cayenne Warbler, *Lath. Syn.*

THIS is as frequent at Cayenne as the preceding: it is nearly of the ſame ſize, but it forms a ſeparate ſpecies, which even includes varieties. The face, the ſides of the head, the anterior part of the back, the wings, and the tail, are of a fine black: the reſt of the plumage is of a fine blue: the bill is blackiſh, and the legs gray.

VARIETIES of the BLUE PITPIT.

1. The bird called by Edwards the *blue manakin:* for the only difference it has from the blue pitpit is, that the throat is black, and the face, as well as the ſides of the head, blue, like the reſt of the body.

2. The

2. The bird figured in the *Planches Enluminées*, No. 669, fig. 1, and denominated the *Blue Pitpit of Cayenne*; the only difference being, that it has no black on the face, or on the ſides of the head.

We muſt obſerve that Briſſon regards the Mexican bird given by Fernandez under the name of *elotototl*, as a blue pitpit:, but we cannot diſcover any foundation for this opinion; ſince Fernandez is the only one who has ſeen that bird, and all that he ſays is, "that the *elotototl* is hardly ſo large as a goldfinch; that it is white or blueiſh, and its tail black; that it inhabits the mountains of Tetzocano; that its fleſh is palatable; that it has no ſong; and, for that reaſon, is not bred in houſes." From this account, it is impoſſible to conclude that this Mexican bird is a blue pitpit more than any other ſpecies.

THE

VARIEGATED PITPIT.

THIRD SPECIES.

Motacilla-Velia, *Linn.* & *Gmel.*
Sylvia-Velia, *Lath. Ind.*
Lufcinia ex Cæruleo & Rubro Varia, *Klein.*
The Red-bellied Warbler, *Lath. Syn.*

THIS bird is found at Surinam and Cayenne. The face is of a fea green: the upper fide of the head, of the neck, and of the back, is of a fine black: the rump is golden green; the throat violet blue; the lower part of the neck and breaft variegated with violet and brown; the reft of the under fide of the body rufous; the fuperior coverts of the tail, and the fmall coverts of the upper fide of the wings, blue; the great coverts and quills of the wings, and thofe of the tail, black, edged with blue: the upper mandible is brown; the lower whitifh: the legs are cinereous. [A]

[A] Specific character of the *Motacilla-Velia*: "It is blue; the belly and rump fulvous."

THE

THE

BLUE-CAPPED PITPIT.

FOURTH SPECIES.

Motacilla Lineata, *Gmel.*
Sylvia Lineata, *Lath. Ind.*
The Blue-ſtriped Warbler, *Lath. Syn.*

THIS is a new ſpecies, and, like the reſt, it is found in Cayenne. We call it the *blue-capped pitpit* *, becauſe it has a ſort of cap of a deep bright blue, which riſes on the face, paſſes over the eyes, and reaches to the middle of the back; only on the crown of the head there is a blue longitudinal ſpot. It is conſpicuous from a white ray that begins at the middle of the breaſt, and extends ſpreading to the under ſide of the tail: the reſt of the under ſide of the body is blue: the bill and legs are black.

* *Pitpit à Coiffe Bleue.*

THE

THE

GUIRA-BERABA.

FIFTH SPECIES.

Motacilla-Guira, *Linn. & Gmel.*
Sylvia Brafilienfis Viridis, *Briff.*
Guira-Guacu-Beraba, *Marcg. Ray, Edw. &c.*
The Guira Warbler, *Lath.*

THIS bird, which Marcgrave has defcribed, appears to me to belong to the pitpits; though his acount is not fufficiently complete to preclude its being ranged with the fig-eaters. It is as large as the goldfinch, which exceeds the ordinary fize of the fig-eaters, and even of the pitpits. The upper fide of its head, its neck, its back, its wings, and its tail, are of a light green: its throat is black: the reft of the under fide of the body and the rump is of a gold yellow: fome quills of the wings are brown at their ends: the bill is ftraight, fharp, and yellow, with a little black on the upper mandible: the legs are brown.

We will obferve that Briffon has confounded this bird with that which Pifo has given under the name of *guira-perea*, though they are certainly

certainly different; for the *guira-perea* of Piſo has its plumage entirely gold colour, except the wings and the tail, which are light green: it is beſides ſpotted like the ſtare on the breaſt and the belly. We need only to compare the two deſcriptions, to ſee evidently that the *guira-perea* of Piſo is not the ſame bird with the *guira-beraba* of Marcgrave, and that they have only the common name of *guira*, but with different epithets. [A]

[A] Specific character of the *Motacilla-Guira*: "It is green; below yellow; its cheeks and throat black, encircled with a yellow line."

THE

YELLOW WREN.

Le Pouillot ou Le Chantre, *Buff.*
Motacilla-Trochilus, *Linn.* & *Gmel.*
Sylvia-Trochilus, *Lath.*
Motacilla Hifpanica, *Haffelq.*
Afilus, *Gefner*, *Briff. Ray*, *Will.* &c.
The Green Wren, *Alb.**

THE three fmalleft of our European birds are the gold-crefted wren, the common wren, and the yellow wren. The latter, though not larger than the reft, is rather longer: it has the fhape, the fize, and figure of a little fig-eater; for the yellow wren might be ranged in that genus, which is already fo numerous, were it not much better to give each fpecies its proper name, which is well known, than to confound it among generic appellations. It might be termed

* The Greek name Οιστρος, and the Latin *Afilus*, fignify a *horfe-fly* or *gad-bee*; and hence were applied to this bird, probably on account of its diminutive fize. In Catalonian, *Xiuxerra*: in Polifh, *Krolic Nieczubaty*: in the Boulonois, *Reatin*: in Provence, *Fifi*: in Burgundy, *Fenerotel* or *Fretillet*: in Lorraine, *Tuit*: in Sologne, *Frelot*, *Frelotte*, *Fouillot*, *Toute-Vive*: in the Orleanois, *Vetti-Vetto*, *Tolitolo*: in Normandy, *Pouillot*, or *Pouliot*.

termed the *little European fig-eater*, and I wonder that ſome nomenclator has not thought of this claſſification. Its French name *pouillot* is evidently derived from the Latin *pullus* or *puſillus*, ſignifying ſmall and feeble.

The yellow wren feeds on flies and other little inſects: its bill is ſlender and tapered, the outſide of a ſhining brown, the inſide and the edges yellow *: the plumage conſiſts of two faint tints of greeniſh gray and yellowiſh white: the firſt ſpreads on the back and the head: a yellowiſh line, riſing from the corner of the bill, paſſes near the eye, and extends to the temple: the quills of the wings are of a dull gray, and, like thoſe of the tail, have their outer edge fringed with greeniſh yellow: the throat is yellowiſh, and there is a ſpot of the ſame colour on each ſide of the breaſt, where the wing reſts: the belly and the ſtomach are white, which is more or leſs daſhed with a weak yellow, according to the age of the bird, or its difference of ſex †: in general, the plumage of the yellow wren reſembles that of the gold-creſted wren, which has only an additional white ſpot on the wing, and a yellow creſt ‡.

The yellow wren reſides in the woods during

* Belon. † Willughby.

‡ Aldrovandus and Belon.

ſummer:

summer: it builds its nest in the heart of the bushes, or in a tuft of thick herbage: the construction is as artful as it is concealed: the outside consists of moss, and the inside is lined with hair or wool: the whole is closely interwoven and covered, and is shaped like a ball, as that of the gold-crested wren, the common wren, and the long-tailed titmouse. It would seem that the voice of nature has directed these four very small birds to the structure of this nest; since their heat, if not guarded and concentrated, would be insufficient for incubation. And this is an additional proof that in all animals the faculty of propagating their species perhaps surpasses the instinct for self-preservation. The female of the yellow wren lays commonly four or five eggs, and sometimes six or seven, of a dirty white, dotted with reddish *. The young ones remain in the nest until they can fly with ease.

In autumn, the yellow wren leaves the woods, and sings in our gardens and vineyards: it seems to repeat the sounds *tuit*, *tuit*, which is

* Willughby and Ray.—" This little bird is much attached to its nest, which it will hardly forsake. A friend of mind told me that, one day having found a nest of this bird, he made it lay thirty eggs one after another, by removing one every day: after which he took pity on the tender dam, and suffered her to hatch." *Salerne.*

the name it receives in ſome provinces*, as in Lorraine, where ſubſiſts no trace of the appellation *chofti* †, beſtowed in the time of Belon; and which, according to him, ſignified *ſinger*, alluding to the variety and continuance of its warble ‡, which laſts during the whole ſpring and ſummer. The ſong has three or four variations, which are moſtly modulated: it begins with a ſlender broken cluck, which is ſucceeded by a ſeries of ſilvery detached ſounds, like the clinking of telling crown pieces: this is probably what Willughby and Albin compare to the ſtridulous voice of graſhoppers. After theſe two notes, very different from each other, the bird ſings its full ſong: it is ſoft, pleaſant, and well ſupported: it laſts during all the ſpring and ſummer; but in the month of Auguſt it gives place to a ſlender whiſtle, *tuit*, *tuit*, which

* In Tuſcany, *Lui;* it pronounces this name with a plaintive voice, ſays Olina, without having any other ſong. This ſeems to ſhew that the yellow wren does not paſs the ſummer in Italy; which is the more probable, as Olina afterwards mentions its being ſeen in winter.

† It has ſtill this name in the foreſt of Orleans. *Salerne.*

‡ This little bird varies infinitely its ſong: it is one of the firſt to announce the return of ſpring. I have heard it ſing more than three weeks before the wild nightingale." *Salerne.*

is nearly the ſame in the red-tail and in the nightingale *.

The yellow wren is extremely active: it inceſſantly flutters briſkly from one branch to another: it darts from its place to catch a fly: it returns and ſearches continually among the leaves, on both ſides, for inſects, which in ſome provinces has given occaſion to the name of *friſker (fretillet, fenerotet)*: it has a ſmall oſcillation of the tail upwards and downwards, but ſlow and regular.

Theſe birds arrive in April, often before the leaves are unfolded. They form flocks of fifteen or twenty during their paſſage; but they immediately ſeparate into pairs. Sometimes, ſhortly after their appearance, they are ſurpriſed by froſts, and drop dead in the roads †.

This delicate little ſpecies is however widely diffuſed. It even viſits Sweden, where Linnæus ſays it inhabits the willow plots ‡. It is known in all the provinces of France: in Burgundy it is called *fenerotet*: in Champagne,

* This is probably what Willughby terms a querulous voice. *Ornithol.* p. 164.

† "This little bird is ſo feeble, that if we caſt a clod at the branch on which it ſits, it will be ſtunned by the ſhake, and tumble off." *Salerne.*

‡ *Fauna Suecica*, No. 236.

fretillet:

fretillet: in Provence, *fifi* *. It is alſo found in Italy †; and the Greeks ſeem to have known it by the term οιστρος ‡ *(a gad-fly)*. It is probable that the *ſmall green wren*, which Edwards tells us was brought from Bengal, is only a variety of the European yellow wren. [A]

[A] Specific character of the Yellow or Willow Wren, *Motacilla-Trochilus:* "It is cinereous greeniſh; the under ſide of its wings and their coverts yellowiſh; its eyebrows yellow." Mr. White aſſerts that there are three ſpecies of the willow wren, which differ in their ſize, and in their note. The yelloweſt bird is conſiderably the largeſt, and is diſtinguiſhed by having its quills tipped with white. It haunts the tops of trees, and makes a ſibilous noiſe like a graſhopper; at intervals it riſes ſinging and ſhivering its wings. But it were raſh to multiply ſpecies. This bird is much ſubject to variety in point of ſize and plumage; and its note muſt evidently depend on the ſeaſon of the year.

* M. Guys. † Aldrovandus.

‡ Ariſtotle (lib. viii. 3) only names the οιστρος between the ὑπολαις and the τυραννος, as one of theſe little birds that feed upon inſects. "Two circumſtances induce us to think that this is the *aſilus:* the firſt, that it would be ſo called in Greece on account of its ſmallneſs; the ſecond, that, as the gad-fly makes a continual noiſe with its wings, ſo this bird ſings almoſt inceſſantly." Belon, *Nat. des Oiſeaux*, p. 344.

THE

GREAT YELLOW WREN.

WE are acquainted with another yellow wren, which is not so small by a fourth part as the preceding, and differs too by its colours: its throat is white, and there is a whitish streak across the eye: a rusty tint, on a whitish ground, covers the breast and belly: the same tint forms a broad fringe on the coverts and quills of the wings, of which the ground is blackish: a mixture of these two colours appears on the back and the head: in other respects, this bird has the same shape with the common yellow wren. It is found in Lorraine, whence it was sent to us: but as we are ignorant of its natural habits, we cannot decide with regard to the identity of these two species.

With respect to the *great yellow wren*, which Brisson reckons, after Willughby, as a variety of the common kind, and which *has double the size*, it is difficult, if the matter is not exaggerated, to conceive that a bird of twice the bulk should belong to the same speciés. Probably Willughby mistook the sedge warbler for the yellow wren, which resembles it much, and is actually twice as large.

THE

N.º 130

FIG 1. THE WREN. FIG 2. THE GOLD CROWNED WREN.

THE

COMMON WREN.

Le Troglodyte, *Buff.*
Motacilla-Troglodytes, *Linn.* & *Gmel.*
Troglodytes, *Geſner*, *Aldrov. Will. Sibb.* &c.
Regulus, *Briſſ.*

THE name *troglodytes* *, which the ancients beſtowed on this little bird, denoted its inhabiting caves or caverns. The moderns have erroneouſly confounded it with the gold-creſted wren: the latter reſorts near our dwellings in winter: it emerges from the heart of buſhes

* In Greek Τροχιλος, from τροχος a top, which comes from τρεχω to run, or whirl; alſo Τρωγλοδυτης, from τρωγλη a cave, or hole, and δυμι to enter: the Romans adopted theſe names, *Trochilus* and *Troglodytes*: in Italian, *Reattino*, *Re di Siepe* (hedge-king): in Tuſcany, *Stricciolo*: in Sicily, *Perchia Chagia*: in German, *Schnee-Koënig*, *Winter-Koënig*, *Zaün-Koënig*, *Thurn-Koënig*, *Meuſe-Koënig*, *Zaun Schlopflin* (the ſnow, winter, hedge, thorn,-king; the hedge-ſlipper): in Swediſh, *Tumling*: in Poliſh, *Krolik*, *Pokrywſka*, *Wolowe Oczko*: in Turkiſh, *Bilbil*: in Provence it is called *Vaque-Petoue*, and *Roi-Bedelet*: in Saintonge, *Roi-Bouti*: in Sologne, *Roi-Berry*: in Poitou, *Quionquion*: in Guienne, *Arrepit*: in Normandy, *Rebetre*: in Anjou, *Berichon*, or *Roi-Bertaud*: in Orleanois, *Ratillon* or *Ratereau*, *Petit-Rat*: in Burgundy, *Fourre-Buiſſon* and *Roi de Froidure*.

 and

and thick boughs, and enters into little lodgements which it makes in the holes of walls. Ariſtotle diſcriminates it by this habit *, and ſelects other features, which it is impoſſible to miſtake; and becauſe of its gold creſt, he terms it little king, or *regulus (roitelet)* †. But the *troglodytes*, or common wren, is ſo different both in its figure and in its œconomy, that the ſame name ſhould never have been applied to it. Yet it is an error of ancient date, perhaps as early as the time of Ariſtotle ‡. Geſner has pointed it out §; but, notwithſtanding his authority, ſupported by Aldrovandus and Willughby, who clearly diſtinguiſh theſe birds ‖, other naturaliſts ſtill perſiſt in confounding them ¶.

The

* "The *trochilus* inhabits orchards and holes; is difficult to be caught, and eluſive." *Ariſt.* lib. ix. 2.

† The *tyrannus* (king), which is not much larger than a locuſt, has a flame-coloured creſt, formed by a ſlight elevation of the plumage: in other reſpects it is beautiful, and ſings ſweetly." Ariſt. *Hiſt. Anim.* lib. viii. 3.

‡ "The *trochilus* is called alſo chief, or king; wherefore the eagle is reported to fight with it." Id. lib. ix. 2.

§ Willughby.

‖ Turner, under the appellation of *trochilus*, deſcribes the common wren; and Ætius gives a very accurate account of it, diſtinguiſhing it judiciouſly from the gold-creſted wren. See Aldrovandus, vol. ii. p. 655.

¶ Olina, Belon, Albin, and Briſſon term it *Regulus*: Friſch and Schwenckfeld, after having named it *troglodytes*, call

The *troglodytes,* then, is that very ſmall bird which appears in the villages and near towns on the approach of winter, and even in the coldeſt weather, having a clear ſprightly little warble, particularly towards evening: it pops out on the top of piles of wood or bundles of faggots, and next moment glides into cover; or if it ventures out on the eaves of a houſe, it quickly hides itſelf under the roof, or in a hole of the wall; when it hops among the heaped branches: its little tail is always cocked. Its flight is ſhort and whirling, and its wings beat ſo briſkly, that their vibrations are not perceptible. Hence the Greeks called it *trochilus*, which is probably the diminutive of *trochus**, a top; and this appellation not only refers to its mode of flying, but correſponds to its round compact form.

The wren is only three inches nine lines long, and its alar extent five inches and an half; its bill ſix lines, and its legs eight: all its plu-

call it likewiſe *Regulus:* but Geſner, Aldrovandus, Johnſton, Willughby, and Sibbald, reject the latter appellation, and adhere to that of *troglodytes*. Klein, Barrère, and Geſner himſelf again apply to the gold-creſted wren the name of *trochilus*, which in Ariſtotle denotes evidently the common wren. Briſſon copies their error.

* "*Trochilus* is derived from *trochus*, becauſe of its top-like ſhape." *Klein*.

mage is interſected tranſverſely with little wavy zones of deep brown and blackiſh on the body and the wings, and even on the head and the tail: the under ſide of its body is mixed with whitiſh and gray: it is the plumage of the woodcock in miniature *. It weighs ſcarcely quarter of an ounce.

This very ſmall bird is almoſt the only one that continues in our climate till the depth of winter; and it alone retains its cheerfulneſs in that dreary ſeaſon: it is always briſk and joyous; and, as Belon ſays, it is conſtantly gay and ſtirring †. Its ſong is loud and clear, and conſiſts of ſhort quick notes, *ſidiriti, ſidiriti:* it is divided by ſtops of five or ſix ſeconds. It is the only light and pleaſant voice that is heard during that ſeaſon, when the ſilence of the inhabitants of the air is never interrupted but by

* I have ſeen children who knew the woodcock call the wren, the firſt time it was ſhewn them, a *young woodcock.*

† The expreſſion uſed is *allègre & vioge*, which, Buffon remarks, has loſt its energy in the French language.

When it ſings, it gives its tail a briſk little motion from right to left. It has twelve quills remarkably tapered; the outermoſt much ſhorter than the next, and this than the third; but the two middle ones are alſo longer than the adjacent one on either ſide: and this property is eaſily perceived, ſince the bird not only cocks its tail, but flies with it ſpread.

the

the disagreeable croaking of the ravens *. The wren sings most when the snow falls †; or in the evening, when the cold threatens to increase the gloom of the night. It thus lives in the out courts and in the wood yards, searching among the faggots, on the bark, under the roofs, in the holes of walls, and even in pits, for chrysalids and dead insects. It frequents too the margins of perennial springs and brooks that never freeze, and shelters itself in the hollow willows. In such lodgments the wrens sometimes gather in numbers ‡: they often come out to drink, and return quickly to their common receptacle. Though familiar, and not disconcerted by near approach, they are difficult to catch: their smallness, and their nimbleness, enable them almost always to elude the eye and the talons of their enemies.

In the spring, the wren lives in the woods, where it builds its nest near the ground, among branches, or even on the turf; sometimes beneath the trunk of a tree, on a rock, or even under the shelving brink of a rivulet; sometimes in the thatched roof of some lone cottage in a wild retreat, and even on the hut of

* Salerne. † Id.

‡ A sportsman told me that he has often found more than twenty collected in the same hole.

the

the charcoal-maker and wooden-ſhoe maker *, who are employed in the foreſts. For the conſtruction the bird collects much moſs, and of that material the outſide is entirely compoſed; but within it is neatly lined with feathers. The neſt is almoſt round, and externally it is ſo bulky and miſhapen as to eſcape the robber's ſearch; for it ſeems only a heap of moſs rolled together by chance: there is only one little narrow aperture made in the ſide. The bird lays nine or ten dirty white eggs †, with a zone dotted with reddiſh at the obtuſe end. It will forſake, if it perceives that they are diſcovered. The young ones leave their lodgment before they are able to fly, and they run like little mice among the buſhes ‡. Sometimes the field-mice poſſeſs themſelves of the neſt; whether that the wren has forſaken it, or that theſe intruders drive away the bird, by deſtroying the hatch §. We have not diſcovered that, in our climate, it breeds a ſecond time during the month of Au-

* In French, *Charbonniers* and *Sabotiers*.

† Schwenckfeld and Aldrovandus. ‡ Geſner.

§ "I found this ſpring in a thorn hedge, about five feet from the ground, a neſt ſhaped like the wren's, built of moſs and wool. I was much ſurpriſed, upon tearing it, to obſerve in it five young field mice. The neſt had been built by the wrens, and the mice had taken poſſeſſion of it." *Note of the Count de Querhoënt.*

guſt.

guſt, as Albertus ſays in Aldrovandus, and as Olina avers to be the caſe in Italy, adding that numbers are ſeen in Rome, and in its vicinity. The ſame author gives directions how to raiſe them, after they are taken from the neſt; but, as Belon obſerves, it will be difficult to ſucceed, for the wren is too delicate *. We have remarked that it is fond of the company of the red breaſts; at leaſt it attends the call with theſe birds: it approaches making a ſhort cry *tirit*, *tirit*, which is of a deeper tone than its ſong, but equally like the ſound of a clock-bell. It is ſo fearleſs and prying, that it even enters the window of the piper's lodge. It flutters and chants in the woods till dark, and, with the redbreaſt and blackbird, it is heard among the lateſt after ſunſet †: it is likewiſe one of the earlieſt awake. It is not prompted, however, by the pleaſures of ſociety; for it prefers retirement in the love ſeaſon, and the males purſue each other hotly ‡.

The ſpecies is extenſively ſpread through Europe.

* "To raiſe it, we muſt keep it warm in the neſt; give it often to eat, but little at a time, ſheep's or calf's heart minced very ſmall, and ſome flies. When it feeds alone, a little corner of the cage ſhould be hemmed in with red cloth, to which the bird may retire at night."

† Turner. ‡ Belon.

Europe. Belon ſays that it is common every where: however, if it endures our winters, it can hardly ſupport the rigours of the North. Linnæus tells us that it is rare in Sweden. The names which it has in different countries ſuffice to diſtinguiſh it: Friſch calls it *king of winter hedges*: Schwenckfeld, *ſnow-king (Schnee-koenig)*: in ſome provinces of France it is termed *chill-king (roi de froidure)*: one of the German names *(Zaun-Schlupfer)* alludes to its gliding into the hedges: and the old Engliſh expreſſion, *dike ſmouler*, mentioned by Geſner, has the ſame import. The Sicilian appellation, *Perchia-chagia*, ſignifies *buſh-borer*. In Orleanois it is called *ratereau*, or *ratillon*, becauſe it runs among the coverts like a young field-mouſe: laſtly, in ſome provinces it is called *ox (bœuf)*, by way of antiphraſis, on account of its extreme ſmallneſs *.

This bird ſeems to have two repreſentatives in the new continent: *the Wren of Buenos-Ayres* and *Wren of Louiſiana*. The firſt is of the ſame ſize and plumage, only its colours are rather more vivid and diſtinct; and it may be regarded as a variety of the European kind. Commerſon, who ſaw it at Buenos Ayres, mentions nothing of its habits, except that it is

* Hebert.

found

found on both banks of the river de la Plata, and that it even enters the veſſels in purſuit of flies. The ſecond is one third larger than the firſt; its breaſt and belly are of a yellowiſh fulvous: there is a ſmall white ray behind the eye: the reſt of the plumage on the head, the back, the wings, and the tail, is of the ſame colour, and marbled, as in the common wren. Father Charlevoix commends the ſong of the Canadian wren, which is probably the ſame with that of Louiſiana. [A]

[A] Specific character of the Common Wren, *Motacilla-Troglodytes*: "It is gray: its wings waved with black and cinereous." Our author, on the authority of Aldrovandus and Geſner, ſays that it lays nine or ten eggs: but Linnæus, Pennant, and Latham agree, that it lays from ten to eighteen. The Wrens continue in Britain the whole year. They are rare in Sweden and Ruſſia, and never penetrate to Siberia.

The North American Wren, mentioned by Charlevoix, appears in the ſtate of New York about May, and breeds in June. It builds in holes of trees; its materials fibres and ſticks, which it lines with hairs and feathers. It lays from ſeven to nine white eggs, with ſtraggling ſpots of red. It retires ſouthwards in Auguſt. It is twice as large as the ordinary wren, and its note is different.

THE

THE

GOLD-CRESTED WREN.

Le Roitelet, *Buff.*
Motacilla-Regulus, *Linn.* & *Gmel.*
Sylvia-Regulus, *Lath.*
Regulus Criſtatus, *Will. Ray,* & *Klein.*
Calendula, *Briſſ.*
Parus Sylvaticus, *Geſner* & *Sibbald.*
The Copped Wren, *Charleton.*
The Gold-crowned Wren, *Edw.* *

THIS is the ſmalleſt of all the European birds. It paſſes through the meſhes of common nets; cages cannot confine it; and, if let looſe in a chamber, the leaſt crack will allow it to eſcape. When it viſits our gardens, and glides among the hedge-rows, how quickly

* In Greek, Τυραννος: in modern Greek, Τετλιγων: in Italian, *Fior Rancio* (marygold flower), *Occhio Bovino* (ox eye), and *Reattino, Reillo, Regillo* (i. e. little king): in Verona, *Capo d'Oro* (gold head): in Genoa, *Boarino della Stella* (ſtar ruſtic): in Bologna, *Papazzino* (little pope): in German, *Gekroëntes* (crowned bird), *Koënigchen* (kingling), *Ochſen Aeuglein* (ox eye), *Holtz-Meiſſe* (wood mouſe): in Swiſs, *Strueſsle:* in Flemiſh, *Konünxken:* in Poliſh, *Krolik, Czubaty, Sikora Leſna:* in Bohemian, *Ztotohtawek:* in Swediſh, *Kongs-Vogel:* in Daniſh, *Fugle-Konge:* in Icelandic, *Rindill.*

does it vaniſh from our ſight! The ſmalleſt leaf is ſufficient to conceal it. If we want to ſhoot it, we muſt uſe very fine ſand; for the ſmalleſt lead ſhot would ſpoil the plumage. When we have ſucceeded in catching it, either with lime-twigs, with the titmouſe trap, or with a very cloſe net, we muſt be careful not to cruſh the delicate bird; and, as it is exceedingly nimble, we are not yet ſure of ſecuring it. It has a ſharp ſhrill cry, like the whiſper of the graſhopper, which is almoſt as large *. Ariſtotle ſays that it ſings agreeably; but probably thoſe who communicated that fact had confounded it with the common wren, eſpecially as the philoſopher himſelf acknowledges that the ſame name was ſometimes applied to both birds. The female lays ſix or ſeven eggs, which are ſcarcely larger than peas: the neſt is formed into a hollow ball, cloſely interwoven with moſs and ſpiders webs, lined with the ſofteſt down, and having an aperture in the ſide. It builds generally in the foreſts; ſometimes on the ivies and elms of our gardens, or on the pines beſide our houſes †.

The

* This ſong is not very harmonious, if Geſner heard it and underſtood it rightly; for he expreſſes it by *zul, zil, zalp*.

† "Lord Trevor found one of theſe neſts in his garden on an

The ſmalleſt inſects are the common food of theſe diminutive birds: in ſummer they catch theſe nimbly on the wing: in winter they ſeek the inſects in their retreats, where they are torpid or dead. They alſo eat the *larvæ*, and all ſorts of worms. They are ſo alert in diſcovering and ſeizing their prey, and at the ſame time ſo great epicures, that ſometimes they continue to ſwallow till they are ſurfeited. During ſummer they feed on ſmall berries and ſeeds, ſuch as thoſe of fennel; and they alſo ſcrape the earth under old willows, where they probably find ſomething nutritious. I never could diſcover ſmall pebbles in their gizzard.

The gold-creſted wrens delight in oaks, elms, tall pines, firs, junipers, &c. In Sileſia, ſays Schwenckfeld, they are ſeen both during the ſummer and the winter, and always in the foreſts. In England they inhabit the mountain woods. In Bavaria and Auſtria they reſort in winter near the towns, where they find reſources for the ſeverity of the ſeaſon. It is ſaid even that they fly in ſmall flocks, which are compoſed not only of their own ſpecies, but of other birds which have the ſame modes of life,

an ivy. Dr. Derham remarks that theſe birds breed every year on the firs before his houſe, at Upminſter, in the county of Eſſex." *Willughby.*

ſuch

ſuch as the creepers, the nuthatches, and the titmice, &c. * On the other hand, Salerne informs us that, in Orleanois, they appear generally in pairs during the winter, and call upon each other when they are ſeparated. It would ſeem, therefore, that they have different habits in different countries; which is not impoſſible, ſince habits depend on circumſtances: but it is more likely that the authors have committed ſome overſight. In Switzerland, it is uncertain whether they continue through the winter: at leaſt, in that country, and in England, they are the laſt to diſappear †. In France, they are oftener ſeen in the autumn and winter than in the ſummer; and there are many provinces where they ſeldom or never breed.

Theſe little birds are very agile and active: they are almoſt continually in motion, fluttering from branch to branch, creeping on the trees, and clinging indifferently in every ſituation, and often hanging by the feet, like the titmice; ferreting in all the cracks of the bark for their diminutive prey, or watching it as it creeps out. In cold weather they lodge in the evergreens, feeding on the ſeed; and often they perch on the ſummit of theſe trees: but in this

* Geſner, Klein, and Cateſby.

† Britiſh Zoology.

habit they appear not to ſhun the preſence of man, ſince, on other occaſions, they ſuffer him to get very near them. In autumn they are fat, and their fleſh is delicate: during that ſeaſon they are commonly caught by means of the call. The public markets of Nuremberg are then well ſtocked with theſe little birds.

The gold-creſted wrens are ſpread not only through Europe, from Sweden to Italy, and probably as far as Spain, but alſo to Bengal: and even in America they inhabit the extent between the Antilles and the north of New England, according to Edwards *. It appears therefore that theſe birds, which viſit the northern countries indeed, but which fly to ſhort diſtances, have migrated from the one continent to the other; and this well-aſcertained fact is a proof of the proximity of the two continents in the high latitudes. If this be admitted, it would follow that the gold-creſted wren, though apparently feeble and delicate, cannot only bear cold, but endure all the viciſſitudes of temperature.

The moſt remarkable part of its plumage is

* It muſt have penetrated much farther, if it be really found in the *Terræ Magellanicæ*, as aſſerted in the *Navigations aux Terres Auſtrales*, t. ii. p. 38. But we cannot infer that the bird here meant is the ſame with the gold-creſted wren.

its beautiful aurora gold crown, bordered with black on each ſide, and which it conceals under the other feathers by the contraction of the muſcles of the head: a white ray, which, paſſing over its eyes, joins the black edging of the crown, and another black ſtreak, in which the eye is placed, give a more marked phyſiognomy: the reſt of the upper ſide of the body, including the ſmall coverts of the wings, is of an olive yellow: all the under ſide, from the baſe of the bill, is light rufous, verging to olive on the ſides: the circumference of the bill is whitiſh, and projects ſome black briſtles: the quills of the wings are brown, edged exteriorly with olive yellow: this border is interrupted, near the third of the quill, by a black ſpot on the ſixth, and more or leſs on the following quills as far as the fifteenth: the middle coverts, and the great coverts next the body, are edged with olive yellow, and tipped with dirty white, which produces two ſpots of the ſame dirty white on each wing: the quills of the tail are dun gray, edged with olive: the ground colour of the feathers is blackiſh, except on the head, at the riſe of the neck, and on the lower part of the thighs; the iris cheſnut, and the legs yellowiſh. In the female the creſt is of a pale yellow, and all the colours of the plumage are more dilute, as uſual.

 The

The Pennſylvanian gold-creſted wren is diſtinguiſhed from this only by ſlight ſhades, which are inſufficient to conſtitute even a variety. The greateſt difference lies in the colour of the legs, which are blackiſh.

Briſſon ſays, that in the gold-creſted wren, the firſt feather of each wing is extremely ſhort: but this is not a quill, its ſhape is different, it is not inſerted in the ſame manner, nor is it deſtined for the ſame uſe: it riſes from the end of a ſort of nail which terminates the bone of the wing; and a ſimilar feather ſprouts from another ſort of nail, at the ſucceeding articulation*.

The gold-creſted wren weighs from ninety-ſix to one hundred and twenty grains.—Total length three inches and a half; the bill five lines; it is black, the edges of the upper mandible are ſcalloped near the tip, and the lower mandible is a little ſhorter: each noſtril is ſeated near the baſe of the bill, and covered by a ſingle feather, which hangs over it with long ſtiff filaments; the *tarſus* ſeven lines and a half; the outer toe connected to the middle one by its two firſt *phalanxes*; the hind nail almoſt double the reſt;

* We may extend this remark to many other ſpecies of birds, which have been ſaid to have the firſt quill of the wing extremely ſhort.

the

the alar extent ſix inches; the tail eighteen lines, conſiſting of twelve quills, of which the two intermediate and the two outer ones are ſhorter than the reſt; ſo that the tail is divided into two equal parts, both tapered: the wings meaſure ſix lines: the body when plucked is not an inch long.

The tongue cartilaginous, terminated by ſmall filaments; the *œſophagus* fifteen lines, dilating and forming a ſmall glandulous ſac before it is inſerted into the gizzard: the gizzard is muſcular, lined with an inadheſive membrane, and covered by the liver: the inteſtinal tube five inches: there is a gall bladder; no *cæcum*. [A]

[A] Specific character of the Gold-creſted Wren, *Motacilla-Regulus*: "Its ſecondary wing-quills yellow on their outer margin, white in the middle; its creſt orange-yellow." It is pretty frequent in England, on the ſummits of lofty trees, particularly oaks. It lays from ſix to eight eggs, which are white, ſprinkled with minute dull-red ſpots. "The gold-creſted wren," ſays the accurate Mr. Pennant, "croſſes annually from the Orknies to the Shetland iſles; "where it breeds, and returns again before winter; a long "flight, of ſixty miles, for ſo ſmall a bird."

VARIETIES OF THE

GOLD-CRESTED WREN.

Motacilla-Calendula, *Linn. & Gmel.*
Sylvia-Calendula, *Lath.*
Calendula-Pennſylvanica, *Briſſ.*

I. THE RUBY-CROWNED WREN. I cannot help conſidering this Pennſylvanian bird as a variety in point of ſize of our gold-creſted wren. In fact, its creſt differs little either in ſhape or colour, being rounder indeed, and of a purer and deeper red, emulating the luſtre of the ruby, and not edged with a black zone. Alſo the upper ſide of the body is olive, which is deeper on the fore parts, and lighter on the rump, without any mixture of yellow: there is a tint of yellow on the lower part of the body, and deeper on the breaſt. But the greateſt difference conſiſts in the ſize, it being larger and heavier than the common gold-creſted wren in the proportion of eleven to eight. For the reſt, the only difference conſiſts in a few ſhades of the plumage: I ſpeak of the dried ſpecimens, for the habits and œconomy of the ruby-crowned wren are entirely unknown; and if ever theſe be diſcovered to be the ſame with thoſe of our gold-creſted wren, the identity of the ſpecies will be completely eſtabliſhed.

In

In the ruby-crowned wrens, the crown is peculiar to the males, and not the leaſt trace of it can be found on the head of the female. However, the plumage is nearly the ſame in both, and their weights are exactly equal.

Total length, four inches and a quarter; the bill, five lines and a half; the alar extent, ſix lines and a half; the *tarſus*, eight lines; the middle toe ſix; the tail eighteen, and conſiſts of twelve quills; it exeeeds the wings about half an inch.

To this variety we may refer the bird which Lebeau found in Louiſiana, in which the back of the head bears a ſort of crimſon crown. The meaſures are indeed a little different, but inſufficient to conſtitute a new variety, and the more ſo, as in other reſpects the birds are analogous, and inhabit the ſame climate.

Total length, four inches and a half; the bill ſix lines; the tail twenty-one lines, and exceeds the wings by eight or nine lines. [A]

[A] Specific character of the Ruby-crowned Wren, *Motacilla-Calendula*: "It is cinereous-greeniſh, a line on its top very yellow: its belly and the under ſide of its wings yellowiſh." It occurs as high as Hudſon's Bay.

II. The Red-headed Wren. This bird was ſeen by Kolben at the Cape of Good

 Hope;

Hope; and though that traveller has not described it completely, we may gather from his account that, firſt, it is a variety of *climate*, ſince it is peculiar to the ſouthern extremity of Africa: ſecondly, it is a variety of *ſize*, ſince, according to Kolben, it is larger than our blue titmouſe: thirdly, it is a variety of *plumage*, for its wings are black, and its legs reddiſh; in which reſpect it differs conſiderably from our gold-creſted wren.

III. This is the place which we ought to aſſign to the bird ſent from Greenland to Muller, under the name of *the ſcarlet-crowned blue titmouſe**, which is all he ſays of it.

* *Zoologiæ Dan. Prodromus*, No. 284. May not this be the *Audua Tytlingr* of the Icelanders?

THE

THE

TITMOUSE-WREN.

Le Roitelet-Meſange, *Buff.*
Sylvia Elata, *Lath. Ind.*

THIS ſpecies, which is found in Cayenne, forms, by its ſhort bill, the intermediate gradation between the gold-creſted wren and the titmice. It is ſtill ſmaller than the gold-creſted wren: it inhabits a hot climate; whereas that bird prefers the more temperate countries, and even appears only in winter. The tit-mouſe-wren lodges in the buſhes in the dry ſavannas, and conſequently near dwellings. It has a jonquil crown on its head, but placed farther back than in the European bird; the reſt of the head is greeniſh brown; the upper ſide of the body, and the two middle quills of its tail, greeniſh; the lateral quills, the ſuperior coverts of the wings, and their middle quills, brown edged with greeniſh, and the great ones brown, without any border; the throat, and the fore part of the neck, light cinereous; the breaſt and the belly greeniſh; the lower belly,

the

the inferior coverts of the tail, and the ſides, dilute yellow.

Total length, three inches and a quarter; the bill four lines (it appears much ſhorter than that of the gold-creſted wren); the *tarſus* ſix lines, and black; the hind nail the ſtrongeſt of all; the tail fourteen lines, conſiſting of twelve equal quills, and exceeds the wings ten lines.

THE

TITMICE*.

Les Mesanges, *Buff.*

THOUGH Aldrovandus has restricted the word *parra* to the gold-crested wren, I conceive that Pliny employed it to signify in general the titmice, and that he regarded this genus as a branch of the family of woodpeckers, which he accounted more extensive than is admitted by the modern naturalists. My reasons are as follow:

1. Pliny says, that the woodpeckers are the only birds which breed in hollow trees †; and it is well known that many species of titmice do the same.

* In Greek the Titmouse is named Αιγιθαλος, Arist. *Hist. Anim.* lib. viii. 3: in Latin, *Parra*; Plin. *Nat. Hist.* lib. x. 33: in modern Latin, *Parus*, *Parix*, *Mesanga.* In Italy it is called *Parula*; and in some districts *Parizola*, *Patascio*, *Parruza*, *Zinzin*, *Orbesina*, *Sparuoczolo*: in Savoy, *Mayenche*: in Germany, *Mayss*, *Meysslin.* The English *Tit-mouse* has the same derivation, and probably, as Ray conjectures, alludes to the bird's nestling in holes of walls like mice.

† *Pullos educant in cavis avium soli.* Lib. x. 18.

2. All

2. All that he ſays in regard to certain woodpeckers, that they climb the trees like cats, that they hang with their heads downward, that they ſeek their food beneath the bark, that they ſtrike it with their bill, &c. agrees equally with the titmice and the woodpeckers*.

3. The account which he gives of other woodpeckers that ſuſpend their neſt from the end of young branches, to prevent any quadruped from approaching it †, will only ſuit ſome kinds of titmice, ſuch as the penduline and the Languedoc, and not at all the woodpeckers properly ſo called.

4. We ſcarce can ſuppoſe, that Pliny never heard of the penduline and Languedoc titmice, ſince one of them at leaſt breeds in Italy; and it is equally improbable that, being acquainted with this fact, he would omit to inſert it in his Natural Hiſtory. But the paſſage alluded to is the only one that applies to theſe birds; and they muſt therefore have been included in the family of woodpeckers.

* *Scandentes in ſubrectum felium modo; illi vero & ſupini percuſſi corticis ſono, pabulum ſubeſſe intelligunt.* Plin. lib. x. 18.

† *Picorum aliquis ſuſpendit in ſurculo (nidum) .... ut nullus accedere poſſit.* Lib. x. 33.

Moreover,

Moreover, the appellation of *parræ** ſeems to have been peculiarly beſtowed on this branch of the woodpeckers; for in the genus of *parræ*, ſays Pliny, there are ſome which form their neſt of dry moſs into a ball, and ſhut it ſo cloſely that the aperture can ſcarce be found. This applies to the common wren, which has been ſometimes confounded with the gold-creſted wren and the titmice. There is another ſpecies which builds in the ſame manner, only employing hemp and flax for the materials; and this is the property of the long-tailed titmouſe. Since the name *parræ* therefore comprehended many ſpecies, and the account of theſe agrees with the qualities of the titmice, it will follow that the genus is really that of the titmice. This idea is the more probable, as the epithet *argatilis*, which is given by Pliny to one of theſe ſpecies, is ſo like the Greek name *aigithalos*, which Ariſtotle applies to the titmice, that we cannot help regarding it the ſame, only ſomewhat altered in tranſcription. Beſides, Pliny uſes the word *aigithalos* in no other part, though he was well acquainted with Ariſtotle's works, and had conſulted them expreſsly in compoſing his tenth book, which

* *In genere parrarum eſt, cui nidus ex muſco arido ita abſoluta perficitur pila, ut inveniri non poſſit aditus.* Plin. lib. x. 33. See Belon, p. 343.

treats

treats of these birds. I may add, that the term *argatilis* has never, as far as I know, been applied by authors to any other bird but the one just mentioned, and there is every reason therefore to conclude that it is a titmouse.

The titmice have also been confounded with the bee-eaters, because they are both *apivorous*: they have been confounded too with the goat-suckers, on account of the resemblance of the Greek names αιγιθαλος* and αιγοθηλης, though Gesner suspects they are distinct in their etymology: besides, the titmice have never been accused of milking the goats.

All the birds of this tribe appear feeble, because they are very small: but they are at the same time lively, active, and bold: they are perpetually in motion; they flutter from tree to tree; they hop from branch to branch; they creep along the bark; they climb the sides of walls; they suspend themselves in all situations, and often their head downwards, in order to dig in every little cranny, and pick out the worms, the insects, or their eggs. They also feed on seeds; but, instead of breaking these in their bill, like the linnets and the goldfinches,

* Αιγιθαλος is commonly reckoned a primitive word, and αιγοθηλης is compounded of αιξ, a goat, and θηλη, a nipple. T.

almost

almoſt all the titmice hold them under their little claws and peck them: they alſo pierce hazel nuts and walnuts, &c.* If a nut be ſuſpended at the end of a thread, they will cling to it, follow the oſcillations, and without quitting their hold they will continue to peck it. It has been obſerved that the muſcles of their neck are very ſtrong, and thoſe of the head thick†; which accounts in part for their manœuvres: their other motions imply great force in the muſcles of the legs and toes.

Moſt of the European titmice occur in our climate at all ſeaſons; but they are never ſo numerous as about the end of autumn; when thoſe which live during the ſummer in the foreſts or on the mountains‡ are driven, by the cold and ſnows, from their retreat, and deſcend, in queſt of food, into the cultivated plains, and near habitations‖. During all the winter months,

* As this exerciſe is rather laborious, and, according to Friſch, brings on blindneſs, it is adviſed to break the nuts and hempſeed, in ſhort, every hard ſubſtance given to them.

† See *Journal de Phyſique—Août* 1776, p. 123, &c.

‡ The long-tailed titmouſe, according to Ariſtotle; the ox-eye, the little blue, the black, and the creſted titmice, according to the moderns.

‖ Some pretend that they retire then into the fir-woods; others aſſert, that they only make tranſient viſits to the ſnowy countries, and advance towards the ſouth. The latter opinion ſeems to be the moſt probable.

and

and even in the beginning of the ſpring, they ſubſiſt on dry ſeeds and on fragments of inſects which they find by ferreting the trees. They alſo crop the opening buds, and eat the caterpillar's eggs, particularly thoſe which are ſeen round the ſmall branches ranged like a ſeries of rings, or the wreaths of a ſpiral. Laſtly, they ſearch in the fields for ſmall dead birds, or ſuch as are exhauſted by diſeaſe, or entangled in ſnares, and, in ſhort, all thoſe incapable of reſiſtance, though of their own ſpecies; they pierce their ſcull, and feed upon the brains. Nor is this cruelty palliated by want; for they are guilty of it even in voleries, where they are abundantly ſupplied. In ſummer they eat not only almonds, walnuts, inſects, &c. but all ſorts of nuts, cheſnuts, beech-maſt, figs, the ſeeds of hemp, of panic, and other ſmall ſeeds*. It is obſerved that thoſe bred in the cage are fond of blood, tainted meat, rancid fat, and tallow melted, or rather burnt, by the flame of a candle. It would ſeem that the ſtate of domeſtication vitiates their taſte.

In general, the titmice, though tainted with ferocity, love the ſociety of their equals, and

* Some pretend that the titmice cannot digeſt the ſeeds of rape or of millet, though theſe be ſoftened by boiling; yet M. de Querhoënt, who raiſed ſome of theſe birds, aſſures me that he fed them only with hempſeed and millet.

unite

unite in numerous flocks: if they are parted by any accident, they mutually call on each other, and soon re-assemble. However, they seem to shun an intimate connection*: judging no doubt of the dispositions of others by their own, they feel that they cannot confide much in them: such is the society of rogues. The unions which they annually form in the spring are of a closer nature, and are very productive. No genus of birds is so prolific as that of the titmice†, and it is the more remarkable the smaller they are. We might suppose that a greater proportion of organic matter enters into their structure, and from this exuberance of life results their fecundity, and also their activity, strength, and courage. No birds attack the owl with such intrepidity; they are ever the first to dart on the nocturnal foe, and they aim constantly at the eyes: their action is attended with a swell of the feathers, and with a rapid succession of violent attitudes and rapid movements, which powerfully mark the bitterness of their rage. When they are caught, they bite keenly the finger of the bird-catcher, strike furiously with their bill, and invite, by their

* *Journal de Physique, Août* 1776, p. 123, &c.

† So well known this fact is in England, that it is usual to call a little prolific woman a titmouse.

 loud

loud ſcreams, the other birds of their ſpecies, which alſo fall into the ſnare, and in their turn decoy others*. Lottinger affirms that, in the mountains of Lorraine, when the weather is foggy, forty or fifty dozens may be caught in a morning† with no apparatus but a call, a ſmall tent, and a cleft ſtick. They may alſo be enſnared with various gins; with the trap‡, with the nooſe, with lime-twigs, or with a ſmall lark-net: or they may be intoxicated, as the ancients practiſed, with meal ſoaked in wine§. Such are the numerous methods of deſtroying theſe ſmall birds, and almoſt all of theſe are ſucceſsfully employed. The reaſon is,

* *Journal de Phyſique, Août* 1776, p. 123.

† According to Friſch, only an hundred are caught in a day by a ſort of ſport in the neighbourhood of Nuremberg. This is performed by means of a triangular lodge, fixed on three large firs that ſerve as columns; each face of this lodge has a ſort of window, in which is ſet a trap, with its decoy-bird. The bird-catcher himſelf keeps in the centre, and ſounds a loud call. *Friſch*, t. 1. claſs 2. This author adds, that ſcarce any are caught in the traps but creſted and long-tailed titmice.

‡ There are cage-traps, and thoſe made with elder and two tiles laid one againſt another, with a head of corn between them; the hurdle, &c.

§ This paſte occaſions giddineſs; they tumble, make efforts to fly, again fall over, and amuſe the ſpectators by the ſtrange variety of their motions and geſtures. Ælianus *de Nat. Anim.* lib. i. 58.

that

that people who keep bees ſuffer much from the titmice, which make great havoc among theſe uſeful inſects, eſpecially when they have young *. Their extreme vivacity drives them into every kind of ſnare, eſpecially on their arrival; for at that time they are very tame, they lodge in the buſhes, and flutter about the roads, allowing one to get near them; but afterwards they gain ſome experience, and become rather more ſhy.

They lay about eighteen or twenty eggs †: ſome depoſit theſe in the holes of trees, which they round and ſmooth with their bill, and faſhion them internally into the proper form; others lay them in ball-ſhaped neſts, which are of a magnitude very diſproportioned to ſo ſmall a bird. We might almoſt ſuppoſe that they previouſly reckon the number of the eggs, and that they anticipate the affection to their expected offspring. Hence the precautions uſed in conſtructing the neſt; the ſolici-

* Others ſay, that winter is the time when theſe birds deſtroy the moſt, becauſe the bees, being then leſs animated, are not ſo formidable with their ſtings, and are more eaſily caught.

† A female, ſays Hebert, that was caught on her eggs, had the ſkin of her belly ſo looſe, that it would have covered the belly entirely, though the bird had been twice as large.

tude which ſome ſpecies diſcover in ſuſpending it from the end of a branch, and the attention in ſelecting the proper materials, ſuch as ſlender graſs, ſmall roots, moſs, thread, hair, wool, cotton, feathers, down, &c. They are able to provide ſubſiſtence for their numerous family, which implies not only indefatigable activity, but much addreſs and ſkill. They are often ſeen returning with caterpillars in their bill. If other birds attack their progeny, they will make an intrepid defence; will dart on the enemy; and courage renders their weakneſs formidable.

All the titmice have white ſpots round the eyes: the outer toe is joined, at its origin, to the middle toe, which is a very little longer than the hind toe: the tongue ſeems truncated, and terminated by filaments: in almoſt all of them are thickly feathered on the rump: all, except the blue one, the head is black or marked with black; in all, except the long-tailed one, the legs are lead-coloured: but what more particularly characteriſes the birds of this family, is, that the bill is not awl-ſhaped, as ſome ſyſtematic writers aſſert, but formed like a ſhort cone, a little flattened on the ſides: it is ſtronger and ſhorter than that of the *fauvettes*, and often ſhaded by the feathers of the forehead, which riſe

riſe and bend forward: their noſtrils are covered with other ſmaller and fixed feathers; and their economy and habits are alſo ſimilar.

It may be worth remarking, that the titmice bear ſome analogy to the ravens the magpies, and the ſhrikes, in regard to the comparative force of their bill and their little talons, in their muſtachoes round the bill, in their appetite for fleſh, in their manner of tearing their food into morſels before they eat; and even, it is ſaid, in their cries, and in their mode of flying: but ſtill we ought not to refer them to the ſame genus, as Kramer has done. We need only to compare theſe birds, to ſee them creeping on the trees, to examine their external ſhape and their proportions, and to reflect on their prodigious fecundity, and we ſhall be convinced that the titmice are widely different from the ravens. Beſides, though the titmice fight among themſelves, and ſometimes devour each other, particularly certain ſpecies which diſcover a violent antipathy*; they ſometimes live on good terms with one another, and even with

* Such are the ox-eye and the cinereous nun. See *Journal de Phyſique, Août* 1776. It is alſo ſaid, that if ſeveral titmice be ſucceſſively put in the ſame cage, the one firſt domeſticated will attack the new comers, will domineer over them, and will endeavour to kill them and ſuck their brains.

birds of other ſpecies; and we may aſſert that they are not radically ſo cruel as the ſhrikes, and only apt to be tranſported by momentary paſſion, in certain circumſtances which are little known. I have witneſſed a caſe, where, far from taking advantage of their ſtrength, when no reſiſtance could have been made, they ſhewed themſelves ſuſceptible of pity and affection. I put two young black titmice, taken from the neſt, in a cage, where was a blue titmouſe: ſhe adopted them, and treated them with the tenderneſs of a mother; ſhared her food with them, and even was attentive to break the ſeeds mixed with it when too hard. I much doubt if a ſhrike would have treated them ſo kindly.

Theſe birds are ſpread through the whole of the ancient continent, from Denmark and Sweden to the Cape of Good Hope, where Kolben ſaw ſix ſpecies, viz. the great titmouſe; the marſh titmouſe; the blue, the long-tailed, the black-headed titmice; and the gold-creſted wren, which he took for a titmouſe. "All theſe kinds ſing pleaſantly," ſays this traveller, "and like Canary finches, with which they mix to form magnificent ſavage concerts *." Our bird-fanciers

* I own that I lay little ſtreſs on this obſervation, in which Kolben, inſtead of relating what he ſaw, ſeems to copy what he read in naturaliſts; only taking the liberty to aſſert that the

ciers pretend that thoſe of Europe alſo ſing well; but this muſt be underſtood of their vernal ſong, which is the muſic of love, and not of the diſagreeable harſh cry which they retain throughout the year, and which has procured them, it is ſaid, the name of *lockſmith**. Theſe connoiſſeurs add, that they can be taught to whiſtle airs: that the young ones, which are caught after they are partly grown, ſucceed better than thoſe fed artificially †: that they ſoon grow tame, and begin to ſing in the courſe of ten or twelve days: laſtly, that they are very ſubject to the cramp, and ought to be kept warm during winter.

Almoſt all the titmice, whether they enjoy the ſtate of liberty, or be confined in a volery, form depoſits for their proviſions. The Viſcount Querhoënt obſerved ſeveral, whoſe wings he had clipped, take in their bill three or four

the titmice ſing like canaries, while theſe authors compare their ſong rather to that of the chaffinches.

* I do not agree with authors on this point; for the name of *lockſmith (ſerrurier)* has been given to the woodpeckers, not becauſe of their cry, but on account of the grating noiſe made by ſtriking the trees with their bill. It ſeems to me more probable that, as the titmice have the ſame habit, they have, for a like reaſon, received the ſame name.

† *Traité du Serin*, p. 51. Every body agrees that the young titmice, taken from the neſt, are difficult to raiſe.

ſeeds of panic, and a ſeed of hemp *, and ſcramble with remarkable agility to the top of the tapeſtry, where they had placed their magazine. But it is obvious, that this inſtinct of ſtoring proceeds from avarice, and not foreſight; at leaſt in the caſe of thoſe which uſually ſpend the ſummer in the mountains, and ſubſiſt during winter in the plains. It has alſo been obſerved, that they conſtantly ſeek the darkeſt ſpots in which to repoſe: they would even ſeem to ſtrive in hollowing out receſſes in the boards or the wall; and theſe attempts are always at a certain height; for they ſeldom reſt on the ground, and never remain long at the bottom of the cage. Hebert took notice of ſome ſpecies which paſſed the night in hollow trees: he perceived them ſeveral times dart briſkly into their lodgment, after they had previouſly glanced round, and, as it were, examined the ground: he tried to drive them out by puſhing a ſtick into the ſame hole by which they entered, but without effect. He ſuppoſes that they return every day to the ſame rooſt; which is the more probable, ſince this alſo contains their little ſtore of proviſions. Theſe birds ſleep ſoundly, with their head concealed be-

* Friſch ſays nearly the ſame thing of the cinereous nun, or marſh titmouſe.

neath

neath their wings, as others. Their flesh is in general lean, bitter, and dry, and consequently very bad food: however, there are some exceptions*.

The largest of all the titmice are, among the European kinds, the great and bearded titmice; and among the foreign kinds, the blue titmouse of India, and the Toupet titmouse of Carolina: each of these weighs near an ounce. The smallest of all are the black-headed titmouse, the long-tailed titmouse, the marsh titmouse, the penduline titmouse, and the crested Carolina titmouse; which exceed not two or three gros.

We shall begin the particular history of the different species, with those of Europe, attending to the characteristic properties of each; and we shall then treat of the foreign kinds. We shall compare these with the European, and mark the analogies that occur; and we shall exclude such as have been inaccurately referred to the genus.

* Gesner says, that they are eaten in Switzerland; but he adds, that they are by no means pleasant food.—— Schwenckfeld alone asserts, that their flesh is neither dry, nor ill-tasted, in autumn and in winter.

THE

GREAT TITMOUSE,

OR

OX-EYE.

La Charbonniere, ou Grosse Mesange, *Buff.*
Parus Major, *All the Naturalists* *.

I KNOW not what induced Belon to assert that "this species does not hang so much from the branches as the others;" for I had occasion to observe one which continually suspended itself from the bars of the upper part of its cage; and, happening to sicken, it still clung to these with its head turned downwards, and continued in this pendent situation during the whole of its illness, and even after its death.

* In Italian, *Parisola Domestica*: in Rome, *Spernuzzola*: in Lombardy, *Parussola*: in Bologna, *Poligola*: in Tuscany, *Cincinpotola*: in German, *Spiegel-meiss* (mirror titmouse), *Brandt-meiss* (fire titmouse), *Kohlmeiss* (coal titmouse): in Dutch, *Coelmaes*: in Swedish, *Talg-Oxe*: in Danish, *Musvit*: in Norwegian, *Kiod-meise*. In different parts of France it is called *Nonnette*, *Moinoton*, *Moineau de Bois*, *Mesange Brûlee*, *Creve Chassis*; which names refer to its dark colour, and its habit of making holes.

I have

I have alſo learned from experience, that the ox-eye kept in the cage ſometimes cleaves the ſkull of the young birds that are preſented to it, and feeds greedily on the brain. Hebert aſcertained nearly the ſame fact by an experiment which he made: he put a red-breaſt in the ſame cage with eight or ten ox-eyes, about nine in the morning; and againſt mid-day the ſcull of the red-breaſt was bored, and the brain entirely eaten. On the other hand, I have ſeen many ox-eyes, and other titmice, which had been caught by means of the call, that lived above a year in the ſame volery, without any act of hoſtility: and, at this very moment, there is an ox-eye which has lived ſix months on good terms with goldfinches and ſiſkins; though one of the ſiſkins was ſick during that period, and, in its feeble ſtate, incapable of reſiſtance, offered an eaſy prey to voracity.

The great titmouſe inhabits the mountains and the valleys, among the buſhes and the copſes, in the vineyards and the foreſts: but M. Lottinger aſſures me that they prefer the mountains. The ordinary cry of the male, which it retains through the whole year, and which is moſt frequent in the evening preceding rain, reſembles the grating of a file, or the grinding of a bolt; and hence, it is ſaid, the appellation of *lockſmith*. In ſpring, however, it aſſumes another

ther modulation, and becomes ſo pleaſant and varied, that we could hardly ſuppoſe it to proceed from the ſame bird. Friſch, Guys, and ſeveral others compare it with that of the finch*; and hence perhaps the reaſon of the name *finch titmouſe*, which has been given to this ſpecies. Olina allows that the ox-eye excels all the other titmice in ſinging, and as a call bird. It is eaſily tamed, and grows ſo familiar as to eat out of the hand: it is dexterous at the little trick of drawing up the pail, and it even lays while in captivity.

When theſe birds enjoy their natural freedom, they begin to pair about the firſt of February: they make their neſt in the hole of a tree, or wall †: but they conſort a long time before they conſtruct it, and they ſelect the ſofteſt

* This bird is kept in the cage in certain countries, ſays Aldrovandus, for the ſake of the pleaſant warble which it has the whole year. On the other hand, Turner ſays, that its vernal ſong is not agreeable, and that the reſt of the year it is mute. According to ſome, it ſeems to ſound *titigu, titigu, titigu*, and in the ſpring *ſtiti, ſtiti*, &c. In general, authors often erect their local obſervations into univerſal axioms; and ſometimes they barely repeat what they have heard from perſons little informed: and hence the contradictions.

† Particularly in the walls of lone houſes, near foreſts: for inſtance, thoſe of charcoal makers (*charbonniers*); whence, according to ſome, the titmouſe has the name of *charbonniere*.

and

and the moſt downy materials. They commonly have eight, ten, or even twelve eggs, with rufous ſpots, chiefly at the large end. The period of incubation exceeds not twelve days: the young brood continue ſeveral days blind: they are ſoon covered with a thin ſlender down, which adheres to the end of the feathers, and drops off as theſe grow. They fly in the ſpace of fifteen days; and it has been obſerved that their growth is more rapid in rainy ſeaſons. After they have once quitted the neſts, they return no more, but perch on the neighbouring trees, and inceſſantly call on each other *: they continue thus in a body, till the approach of ſpring invites them to pair. The neſtlings are found till the end of June, which ſhews that the ox-eyes have ſeveral hatches. Some ſay that they have three: but is it not when they are diſturbed in the firſt hatch that they begin a ſecond, &c.? Before the firſt moulting the male may be diſtinguiſhed, ſince he is larger, and of a hotter temper. In the ſpace of ſix months the young are all full grown, and four months after moulting they are fit for breeding. According to Olina, theſe

* It is perhaps an effect of this early habit, that the titmice run ſo nimbly when they hear the voice of their fellows.

birds live only five years; and others mention that age as the time when they begin to be afflicted with defluxion of the eyes, the cramp, &c. But they lose their activity without losing the harshness of their character, which is aggravated by their infirmities *. Linnæus says, that in Sweden they lodge on alders, and that in summer they are very common in Spain.

The great titmouse has on its head a sort of cowl of a bright glossy black, which before and behind descends to the middle of the neck, and has on each side a large white spot almost triangular: below this cowl rises before a long narrow black bar, which extends across the middle of the breast and belly to the extremity of the inferior coverts of the tail: these are white, and also those of the lower belly: the rest of the under surface of the body, as far as the black on the throat, is of a light yellow: an olive green prevails on the upper side of the body, but becomes yellow, and even white, as it approaches the lower edge of the cowl: it grows duskier, on the contrary, at the opposite side, and changes into a blue ash colour on the rump and the superior coverts of the tail: the two first quills of the wing are of a brown cinereous, without edgings: the rest of the great quills

* *Journal de Physique, Août* 1776.

are

are bordered with blue aſh, and the middle ones with olive green, which aſſumes a yellow tinge on the four laſt: the wings have a tranſverſe ray of yellowiſh white: all that appears of the quills of the tail is blueiſh cinereous, except the outermoſt, which is edged with white, and the next, which is tipped with the ſame colour: the ground of the black feathers is black, and of the white ones white: that of the yellow ones is blackiſh, and that of the olive ones cinereous. The bird weighs about an ounce.

Total length ſix inches; the bill ſix lines and an half; the two mandibles equal; the upper one has no ſcalloping: the *tarſus* nine lines; the hind nail the ſtrongeſt; the alar extent eight inches and an half; the tail two inches and an half, ſomewhat forked, conſiſting of twelve quills, and exceeds the wings eighteen lines.

The tongue is not fixed and immoveable, as ſome have ſuppoſed *: the bird can puſh it forward and raiſe it parallel to itſelf, with a moderate declination to the right and left; and conſequently it is ſuſceptible of all the motions that can be compounded of theſe three: it is truncated at the end, and terminated with three

* *Journal de Phyſique,, Août* 1776.

or four filaments. Frifch fuppofes, that the great titmoufe ufes thefe to tafte its food before eating.

The *œfophagus* two inches and an half, forming a fmall glandulous fac before its infertion into the gizzard, which is mufcular, and lined with a wrinkled inadhefive membrane. I found in it fmall black feeds, but not a fingle pebble: the inteftines fix inches four lines; two veftiges of a *cæcum*; a gall bladder. [A]

[A] Specific character of the Ox-eye, *Parus Major*: "Its head black, its temples white, its nape yellow." A variety of this bird was killed at Feverfham in Kent: its colours were in general more obfcure, its bill was very long, and its mandibles bent at the tips. The ox-eye occurs in the northern extremities of Europe, of Afia, and even of America: it is a permanent fettler. Its egg is white, with numerous rufous fpots.

No 131

THE COLEMOUSE.

THE

COLEMOUSE*.

La Petite Charbonniere.

Parus Ater, *Linn. Gmel. Kram. Frif. Ray & Will.*
Parus Atricapillus, *Briff.*
Parus Sylvaticus, *Klein.*
Parus Carbonarius, *Barrere.*

THE name of black-head (*atricapilla, μελαγκορυφος*) has been applied to feveral birds, fuch as the black-cap, the bulfinch, &c. But the black-head of Ariftotle appears to be a titmoufe; fince, according to him, it lays feventeen, or even twenty-one eggs, and has befides all the properties of the titmice; fuch as neftling in trees, feeding on infects, having a truncated tongue, &c. What this author adds from report, and which Pliny confidently repeats, that the eggs are always odd, is founded on the notion of certain myfterious properties of numbers, efpecially of the odd ones, which have ever been fuppofed to influence the phænomena of nature.

* In Italian, *Cingallegra*: In German, *Tannenmeife, Kleine Kohlmeife, Hunds-meife*: in Polifh, *Sikora Czarna Mnicyffa.*

The colemouſe differs from the ox-eye, not only in regard to ſize, being no more than the third or fourth of the weight, but alſo by the colours of its plumage; as will appear by comparing the deſcriptions. Friſch ſays, that in Germany it inhabits the pine foreſts; but in Sweden it prefers the alders, according to Linnæus. It it the leaſt timorous of all the titmice: not only the young ones flock to the voice of another titmouſe, and are decoyed by means of the call; even the adults, which have been caught ſeveral times and have fortunately eſcaped, are as eaſily enſnared again in the ſame gins. However, theſe birds diſcover as much or even more ſagacity in many actions which concern their own preſervation, or that of their brood; and perhaps the courage which they poſſeſs extinguiſhes alike miſtruſt and fear.

The colemouſe lives in the woods, eſpecially thoſe which contain firs and other evergreen trees, in vineyards, and gardens: it creeps and runs on the trees like the other titmice. Next to the long-tailed one, it is the ſmalleſt of all; it weighs only two gros: it has a ſort of black cowl, terminated with white on the back of the head, and marked below the eyes with the ſame colour: the upper ſide of the head is cinereous,

nereous, the under dirty white: there are two transſverſe white ſpots on the wings; the quills of the tail and of the wings, brown aſh colour, edged with gray: the bill black, and the legs lead-coloured.

Total length, four inches and a quarter; the bill, four lines and two-thirds; the *tarſus*, ſeven lines; the hind nail is the ſtrongeſt; the lateral ones are proportionally longer than in the ox-eye: the alar extent, ſix inches and three quarters; the tail, twenty lines, and rather forked, conſiſting of twelve quills; it exceeds the wings ten lines.

Mœhring has obſerved that, in this ſpecies, the end of the tongue is truncated only at the edges, from each of which a filament projects, and that the intermediate ſpace is entire, and riſes almoſt perpendicular. [A]

[A] Specific character of the Colemouſe, or Coalmouſe, *Parus Ater*: "Its head black, its back cinereous, the back of the head and the breaſt white." It inhabits as far north as Siberia, where it continues even through the winter: its egg is whitiſh, with ſmall reddiſh ſpots.

VARIETIES OF THE COLEMOUSE.

I.

THE MARSH TITMOUSE, OR BLACKCAP.

La Nonnette Cendrée, *Buff.*
Parus Paluſtris, *Linn. Gmel. Briſſ. &c.* *

I KNOW that many naturaliſts regard this bird as diſtinguiſhed from the preceding by conſiderable differences. Willughby ſays that it is larger, its tail longer, and that it has leſs black under its throat; that the white of the lower ſide of its body is purer, and that it has no white on the back of the head, or on the wings. But if we conſider that moſt of theſe diſtinctions are not permanent, particularly the ſpot on the back of the head †, though it is reckoned among the ſpecific characters of the colemouſe; if we conſider that the ſame name (*charbonniere*) *collier* has been applied to both, and the epithet of *marſh*, which is commonly

* In Italian, *Parolozino Paluſtre:* in German, *Gartenmeiſe* (garden-titmouſe), *Bien-Meiſe* (bee-titmouſe): in Swediſh, *En-Tita Tomlinge:* in Poliſh, *Sikora Popielata.*

† A colemouſe obſerved by the authors of the Britiſh Zoology wanted this ſpot; and M. Lottinger aſſures me, that if the marſh titmouſe had this ſpot on the back of the head it would differ not from the colemouſe.

beſtowed

beſtowed on the blackcap, would alſo ſuit the preceding ſpecies, ſince, according o Linnæus, it lodges in alders, which grow in wet ſituations: laſtly, if we conſider the numerous points of analogy that ſubſiſt between the two ſpecies, their haunts, their ſize, their breadth, the ſameneſs of their colours, and their nearly ſimilar diſtribution, we ſhall be convinced that it is only a variety of the colemouſe*.

The marſh titmouſe lives more in the woods than in the vineyards and gardens, feeding on ſmall ſeeds, preying on waſps, bees and graſhoppers, and forming ſtores of hemp ſeeds, when there is occaſion, carrying ſeveral at once in its bill to place in the depoſit, and conſuming them afterwards at leiſure. Its manner of eating undoubtedly gives it this foreſight: it requires time and a convenient place to pierce each ſeed with its bill; and if it had not collected proviſions, it would often be reduced to want.

The marſh titmouſe is found in Sweden, and even in Norway, in the foreſts which ſkirt the Danube, in Lorraine, in Italy, &c. Salerne ſays that it is unknown in the Orleanois, in the neighbourhood of Paris, and in Normandy. It is fond of lodging among the alders, the ſallows, and of haunting wet ſpots. It is a ſolitary bird, which continues in the country

 through

through the whole year; it can hardly be bred in the cage. A neſt was brought to me which was found in a hollow apple-tree in the midſt of a little clump, not far from a river: it conſiſted of a little moſs laid in the bottom of the hole. The young ones, which were already able to fly, were rather browner than the parent, but their legs were of a lighter lead colour; no ſcaloping on the edges of the bill, of which the two mandibles were very equal. What was remarkable, the gizzard of the young ones was larger than that of the adults in the ratio of five to three; the inteſtinal tube was alſo proportionally longer; but neither of them had a gall bladder or the leaſt veſtige of a *cæcum.* I found in the gizzard of the parent ſome fragments of inſects and a grain of dry earth, and in that of the young ones ſeveral little pebbles.

The marſh titmouſe is rather larger than the colemouſe, for it weighs three gros. I need not deſcribe its plumage; it will be ſufficient to mark the chief differences.

Total length, four inches and one third; the bill, four lines; the *tarſus*, ſeven lines; the alar extent, ſeven inches; the tail two inches, conſiſting of twelve quills, and projects twelve lines beyond the wings. [A]

[A] Specific character of the Marſh Titmouſe, *Parus Paluſtris*: "Its head black, its back cinereous, its temples white."

M. Le

M. Le Beau brought from Louiſiana a titmouſe, which reſembles much the preſent; only it wants the white on the back of the head, and the two ſtreaks of the ſame colour on the wings; alſo the black mark on the throat was larger, and in general the colours of the plumage rather deeper, except that in the female the head was of a ruſty gray, nearly like the upper ſide of the body, but ſtill darker.

Total length, four inches and a half; the *tarſus*, ſeven or eight lines; the hind nail the ſtrongeſt; the tail twenty-one lines, rather tapered (which is another diſtinctive character), and exceeding the wings about nine lines.

II.

THE CANADA-TITMOUSE.

Parus Atricapillus, *Linn.* & *Gmel.*
Parus Canadenſis Atricapillus, *Briſſ.*

THE black-headed titmouſe of Canada bears a great reſemblance to the colemouſe: it has nearly the ſame proportions and the ſame plumage; the head and throat black; the under ſide of the body white; the upper ſide dark

 cine-

cinereous, which, towards the rump, grows more dilute, and, on the ſuperior coverts of the tail, runs into a dirty white: the two intermediate quills of the tail are cinereous, like the back; the lateral ones alſo cinereous, but edged with white gray; thoſe of the wings brown, edged with the ſame white gray; their great ſuperior coverts brown, edged with gray; the bill black, and the legs blackiſh.

Total length, four inches and a half; the bill, five lines; the *tarſus*, ſeven lines and a half; the alar extent, ſeven inches and a half; the tail, twenty-ſix lines, conſiſting of twelve equal quills, and exceeds the wings an inch.

Since the titmice frequent the northern countries, it is not ſtrange that we ſhould find in America varieties of the European ſpecies. [A]

[A] Specific character of the *Parus Atricapillus*: "Its cap and throat are black; its body cinereous, below white."

III.

IF the white-throat of Willughby be not a fauvette (*ſpipola*), as he ſuppoſes, but a titmouſe, as Briſſon reckons it*; I ſhould range

* Parus Cinereus, *Briſſ.*

it

it with the marſh titmouſe, and conſequently with the colemouſe. Its head is deep cinereous; all the upper ſide of the body-ruſty cinereous; the under ſide white, tinged with red in the male, except the origin of the neck, which in ſome ſubjects is pure white, and in others has a cinereous tinge, as well as the fore ſide of the neck and breaſt: the firſt quill of the wing is edged with white, the laſt ones with rufous; the quills of the tail black, edged with lighter colour, except the outermoſt, which is white, though not in all ſubjects: the bill is black, yellow internally; the lower mandible whitiſh in ſome ſubjects; the legs ſometimes yellowiſh brown, and ſometimes lead coloured.

The white-throat is found in England during the ſummer: it viſits the gardens, lives on inſects, makes its neſt in the buſhes near the ground (and not in the holes of trees like the titmice), and lines it with hair: it lays five eggs dotted with black, on a greeniſh light brown ground. It is nearly as large as the marſh titmouſe.

Total length, from five inches and three quarters to ſix inches; the hind toe the ſtrongeſt; the two lateral ones equal, exceedingly ſmall, and connected to the mid one, the outer by by its firft phalanx, the inner by a membrane, which is uncommon in birds of this kind; the alar extent, eight inches; the tail two inches and a half, confifting of twelve quills, rather tapered; it exceeds the wings fixteen or feventeen lines*.

IV.

I HAVE at prefent before me a bird which the Marquis de Piolenc fent from Savoy under the name of *creeper*, but which muft be referred to the fame fpecies. Its head is variegated with black and cinereous gray; all the reft of the upper part, including the two middle quills of the tail, of the fame gray; the outer quill blackifh at the bafe, gray at the end, and croffed near its middle by a white fpot; the next quill marked with the fame colour, but on its infide only; the third in the fame way, but nearer the end,

* I have feen in cabinets a bird, whofe plumage refembled remarkably that of this titmoufe, but which differed by its proportions. Its total length was five inches and a half; its *tarfus*, ten lines; its tail, twenty-nine lines, exceeding the wings only by an inch: but the moft remarkable circumftance that difcriminated it was its bill, feven lines long, and three lines thick at the bafe.

fo

ſo that the white always contracts, and the black extends ſo much farther: it gains ſtill more in the fourth and fifth quill, which have no white at all, but are tipt with aſh-gray, as in the preceding: the quills of the wings are blackiſh; the middle ones bordered with aſh-gray; the great ones with dirty gray: each wing has a longitudinal, or rather a yellowiſh white ſtreak: the throat is white, and alſo the anterior margin of the wing; the fore part of the neck and all the lower part light rufous: the inferior coverts of the wings, the neareſt the body, are ruſty, the ſucceeding black, and the longeſt white: the upper mandible is black, except the ridge, which is whitiſh, and ſo is the lower mandible: laſtly, the legs are yellowiſh brown.

Total length, five inches one third; the bill, ſix lines and a half; the *tarſus*, eight lines; the hind toe as long and thicker than the mid one, and its nail the ſtrongeſt; the alar extent, ſeven inches and three quarters; the tail eighteen lines, conſiſting of twelve quills, rather unequal, and ſhorter in the middle; it exceeds the wings ten lines.

THE

BLUE TITMOUSE.

La Mesange Bleue.

Parus Cæruleus, *Linn. Gmel.* & *Briss.* &c.
The Nun, *Charleton**.

FEW birds are so well known as this; because few are so common, so easily caught, or so distinguished by the colours of their plumage. Blue predominates on the upper side, yellow on the lower; and a nice distribution of black and white discriminates and heightens the different hues, which are also diversified by a variety of different shades. The blue titmouse is the better known on account of its pernicious visits to our gardens, where it plucks the blossoms from the fruit-trees: it even dexterously employs its little claws to detach the ripe fruit from the branch, which it afterwards carries to its deposit. But it does not subsist wholly in this way; it has the same

* In Italian, *Paro-zolino*, *Fratino:* in Spanish, *Milchiero:* in Portuguese, *Chamaris Alionine:* in German, *Blaw meise* (blue titmouse), *Meel-meise* (meal titmouse): in Swedish, *Blao-mees:* in Norwegian and Danish, *Blaa-meise:* in Polish, *Sikora Modra.*

pro-

propenſity to fleſh with other titmice, and it picks ſo clean the bodies of the little birds which it maſters, that Klein propoſes to employ it for preparing their ſkeletons *. It alſo diſtinguiſhes itſelf above all the reſt by its rancour againſt the owl. The Viſcount de Querhoënt obſerves, that it does not always ſplit the ſeeds of hemp like the other titmice, but bruiſes them in its bill, like the canaries and linnets: he adds, that it ſhews more foreſight than the reſt, ſince it ſelects for its winter haunt a warmer ſite, and one of more difficult acceſs; commonly a hollow tree, or the crevice of a wall.

The female alſo builds in holes, and is not ſparing of feathers; ſhe lays in the month of April a great many ſmall white eggs: I have counted from eight to ſeventeen in the ſame neſt; others have found even twenty-one. I am aſſured that there is only one hatch, unleſs the incubation is diſturbed: a ſingle egg broken, or even handled, will occaſion the whole to be forſaken. But after the young are excluded, the mother diſcovers a ſtrong attachment, and defends them with courage; ſhe is inflamed, and whiſtles a

* He adviſes us previouſly to remove moſt of the fleſh and the brain of the bird, whoſe ſkeleton we want to have prepared.

threat-

threatening air when difturbed in her prifon. The cock feems to repofe more at eafe, and ofteneft clung to the top of the cage. Befides the difagreeable grinding, it has a flender but varied chirp, which has been fuppofed to bear fome refemblance to that of the chaffinch.

Frifch alleges that the blue titmoufe will not live in the cage, and cannot therefore be ufed as a call-bird. I have feen fome, however, that were kept many months in confinement, and died only of exceffive fat.

Schwenckfeld tells us, that in Silefia this titmoufe is found at all feafons on the mountains. With us, it prefers the woods, efpecially in fummer, and next to thefe the vineyards, the gardens, &c. Lottinger fays, that it travels in company with the ox-eye; but the fociety between petulant and cruel animals muft be turbulent and temporary. It is faid that its young continue longer together than that of the other fpecies *.

The blue titmoufe is very little, fince it weighs only three gros: but Belon, Klein, and the traveller Kolben, ought not to have reprefented it as the fmalleft of the titmice. The hen is rather fmaller than the cock, and has lefs

* *Journal de Phyfique, Août* 1776.

blue

blue on the head; and this blue, as well as the yellow of the under ſide of the body, is not ſo bright: what is white in the parents, is only yellowiſh in the new-fledged brood: what is blue in the former, is aſh brown in the latter; but their wing quills are proportionally as large as thoſe of the adults.

Total length, four inches and a half; the bill, four lines and a half; the two mandibles equal, and without any indenting; the tongue truncated, terminated with filaments, ſome of which are commonly broken; the *tarſus*, ſix lines and a half; the legs are exceedingly thick, the hind nail the ſtrongeſt; the alar extent, ſeven inches; the tail twenty-five lines, and projecting twelve beyond the wings; each of its halves is tapered, and compoſed of ſix quills. The young ones, of which I diſſected a pretty large number in May, had all of them rather a ſmaller gizzard than that of their dam, but a longer inteſtinal canal. There were two ſlight veſtiges of *cæca*, and no gall bladder. [A]

[A] Specific character of the Blue Titmouſe, *Parus Cæruleus*: "Its wing quills are blueiſh; the firſt white at their outer margin, the front white, the top blue."

THE

THE

BEARDED TITMOUSE.

La Moustache.

Parus Biarmicus, *Linn. Gmel.* & *Frif.*
Parus Barbatus, *Briff.* & *Klein.*
Lanius Biarmicus, *Brun.*
Pendulus, *Kram.*
The Leaft Butcher-bird, *Edw.* *

I CANNOT be pofitive whether this bird is really found in India, as Frifch feems to hint; but it appears to be very common in Denmark, and is now fpreading through England. Edwards mentions feveral cocks and hens that were killed in the neighbourhood of London, and fo little known that they had no name. The Countefs of Albemarle brought a cage full of them from Denmark †; and fome of thefe doubtlefs efcaped, and founded a colony in England. But whence came thofe which

* In German, *Spitz-bartiger* (bearded fparrow).

† Since they are fo common in Denmark, I am furprifed that the name occurs not in Muller's *Prodromus Zoologiæ Daniçæ*.

Albin

No. 132

FIG I. THE MALE BEARDED TITMOUSE FIG. 2. THE FEMALE.

Albin ſays were reported to inhabit the counties of Eſſex and Lincoln, and always among the fens?

It is to be wiſhed that the habits of theſe birds were better known. Their hiſtory would be curious, if we may judge from what is related, that when they reſt the male ſpreads his wings over his mate; and this attention, were it well authenticated, muſt imply many other intereſting particulars with regard to incubation.

The moſt characteriſtic feature of the male is a black mark, very nearly triangular, on each ſide of the head: the baſe of this inverted triangle riſes a little above the eyes, and its vertex is turned downwards, and falls on the neck nine or ten lines from the baſe: theſe two black marks, which have pretty long feathers, bear ſome reſemblance to whiſkers, and hence the names of the bird in different countries. Friſch ſuppoſes that it is analogous to the canary, and that the two ſpecies would intermix; but adds, that the bearded titmice are too rare for making the neceſſary experiments. This opinion of Friſch is inconſiſtent with that of Edwards and Linnæus, who ſuppoſe it to reſemble the ſhrike. But though theſe views be oppoſite, they agree in one circumſtance, that the bill of the bearded titmouſe is larger than or-

 dinary

dinary in titmice. On the other hand, Lottinger affirms that it breeds in the holes of trees, and often consorts with the long-tailed titmouse; which, joined to the family likeness and other resemblances in size, exterior figure, mien, and habits, rank it with the titmice.

The head of the male is pearl gray; the throat, and the fore part of the neck, of a silvery white; the breast of a sullied white tinged with gray in some subjects, and rose-colour in others; the rest of the under side of the body rusty; the inferior coverts of the tail black; those of the wings yellowish white; the upper side of the body light rufous; the anterior edge of the wings white; the small superior coverts blackish; the great ones edged with rufous, the middle ones with the same, edged interiorly with lighter rufous; the great quills edged with white externally; those of the tail entirely rufous, except the outermost, which is blackish at the base, and of a rufous ash colour near its extremity; the iris orange; the bill yellowish, and the legs brown.

In the female there is no red tinge under the body, nor black marks on the sides of the head, which is brown, and also the inferior coverts of the tail, of which the lateral quills are blackish tipt with white. The female is also rather smaller than the male.

 Total

Total length of the laſt, ſix inches and a quarter; the bill, leſs than ſix lines; the upper mandible a little hooked, but without any indenting, according to Edwards himſelf, which is very different from a ſhrike; the *tarſus*, eight lines and a half; the alar extent, ſix inches and a half; the tail thirty-ſix lines, conſiſting of twelve tapered quills, ſo that the two exterior ones are only half as long as the two middle ones; it exceeds the wings twenty-ſeven lines. [A]

[A] Specific character of the Bearded Titmouſe, *Parus Biarmicus*: "Its head hoary, its tail longer than its body, its head bearded."

THE

PENDULINE TITMOUSE.

Le Remiz, *Buff.*
Parus Pendulinus, *Linn. Gmel. Georg. Borowsk.*
Parus Polonicus, sive Pendulinus, *Briss.*
Parus Lithuanicus, *Klein.*
The Mountain Titmouse, *Alb.* *

EDWARDS suspects this bird, which is figured in the work of Albin, to be the same with the bearded titmouse; but this idea seems contradicted by several circumstances: 1. The figures referred to, though accurate, differ considerably. 2. According to Albin, the bearded titmouse weighs more than nine gros, and he represents the mountain titmouse as equal to the

* In Italian, *Pendolino*: in German, *Weiden-meise* (willow titmouse), *Rohr-spatz* (reed sparrow), *Persianischer-spatz*, *Turquisher-spatz*: in Polish, *Remez*, *Remis*, *Remiz*, *Remizawy*, *Ptak*, *Remicz*, *Remits'ch*, *Remisch*, *Romisch*: in Russian, *Remessof*: in Hungarian, *Maundicek*.

The name *Picus nidum suspendens* (hang-nest woodpecker), which Aldrovandus gives from Pliny, would agree much better with the penduline titmouse, or remiz. Some have reckoned a land remiz, and a water remiz; but the latter is probably the reed-bunting. Lastly, the anonymous author of a Memoir inserted in the *Journal de Physique* for August 1776, gives the remiz the name of *marsh titmouse*.

blue

blue titmouſe, which is only three gros. 3. The plumage is different; and particularly the black bar on the ſide of the head has a very different poſition in each. 4. The climate alſo differs: for Albin makes ſome of the counties in England the ordinary haunt of the bearded titmouſe; but regards the penduline titmouſe as an inhabitant of Germany and Italy. Nor do Kramer and Linnæus appear to judge better in conſidering theſe two birds as diſtinguiſhed only by their ſex; and I cannot, with Edwards and Linnæus, perceive any ſtrong reſemblance between theſe and the ſhrikes. It is true that the ſhrikes have a black ſtripe on the eyes, and that the penduline titmouſe interweaves its neſt; but the materials are different, and alſo the manner of faſtening the neſt, the ſhape of the bill, the claws, the food, the ſize, proportions, ſtrength, geſture, &c. It would ſeem that neither Edwards, nor any of the naturaliſts who have adopted his opinion, had ever ſeen this bird.

The moſt curious fact in the hiſtory of theſe birds is the exquiſite art diſplayed in the conſtruction of their neſt. They employ the light down found on the buds of the willow, the poplar, the aſpen, the juncago; in thiſtles, dandelions, flea-banes, cats' tails, &c. With their bill they entwine this filamentous ſubſtance, and form a thick cloſe web, almoſt like cloth: this they fortify externally with fibres and ſmall roots, which penetrate into the texture, and in ſome meaſure form the baſis of the neſt. They line the inſide with the ſame down, but not woven, that their young may lie ſoft *: they ſhut it above to confine the warmth, and they ſuſpend it with hemp, with nettles, &c. from the cleft of a ſmall pliant branch, over running water, that it may rock more gently, aſſiſted by the ſpring of the branch. In this ſituation the brood are well ſupplied with inſects, which conſtitute their chief food †; and they are protected from the rats, the lizards, the adders, and other reptiles, which are always the moſt dangerous: and I am convinced that their conduct really proceeds from foreſight; for they are naturally crafty, and, according to Monti and Titius ‡, they can never be caught in ſnares; as the ſame circumſtance has been remarked in the bonanas

* Sometimes the down, or cotton-like ſubſtance, is rolled into little pellets, which make the inſide of the neſt neither ſo ſoft nor ſo pleaſant.

† M. Monti found in the ſtomach of theſe birds inſects almoſt digeſted, and nothing elſe.

‡ They are ſometimes, adds Titius, ſurpriſed in the neſt at ſun-ſet, or in dark miſty weather.

and

and caſſiques of the N w World, in the grosbeaks of Abyſſinia, and in other birds which hang their neſts from the end of a branch. That of the penduline titmouſe reſembles ſometimes a bag, ſometimes a ſhut purſe, ſometimes a flattened bagpipe, &c.* The aperture is made in the ſide, and almoſt always turned towards the water, and placed ſometimes higher, ſometimes lower: it is nearly round, and only an inch and an half in diameter, or even leſs, and com-

* Cajetan Monti has cauſed one to be engraved, and Daniel Titius two. Theſe three neſts differ not only from each other, but from that figured by Bonanni, both in ſize and form. The largeſt of all (Titius, *pl.* 2.) was ſeven inches long, and four and an half wide: it was ſuſpended at the fork of a ſmall branch with hemp and flax. The leaſt (*pl.* 1.) was five inches and an half long, of the ſame breadth at its upper part, and terminating in an obtuſe point: this, according to Titius, is the uſual form. That of Monti was pointed above and below. Titius ſuſpects that the penduline titmice only make a rude eſſay in conſtructing their firſt neſt, and that the ſides are then thin, and the texture quite looſe; but that they improve at each ſubſequent hatch; and, as their miſtruſt grows upon them, they add firmer coats on the outſide, and ſofter ones within: hence the differences obſerved in the form and bulk of theſe neſts. About the end of December 1691, near Breſlaw, was found a female ſiſkin in one of theſe ſame neſts, with a young one, and three eggs not yet hatched. This proves that the neſts of the penduline titmice ſubſiſt from one year to another. Titius adds, that we need not wonder to ſee the ſiſkin hatching in winter, ſince the croſs-bills do the ſame.

monly ſurrounded by a brim more or leſs protuberant *; though this is ſometimes wanting. The female lays only four or five eggs, which falls much ſhort of the ordinary fecundity of the titmice; but in its port, its voice, its bill, and in the principal attributes, the penduline reſembles the others. Theſe eggs are as white as ſnow, the ſhell extremely thin, and they are almoſt tranſparent. The bird has generally two hatches annually; the firſt in April or May, and the ſecond in the month of Auguſt. There is little probability that it makes a third.

Theſe neſts of the penduline titmouſe are ſeen in the fens near Bologna, in thoſe of Tuſcany, on the lake Thraſymene, and are exactly

* Aldrovandus has given a figure of this neſt, which he imagined to be the long-tailed titmouſe, though he well knew that the bird was called *pendulino*. *Ornithol.* t. ii. p. 718. Two of theſe neſts ſeem glued together, and remind us of what Rzacynſki ſays of the neſts with double apertures, found in Pokutia, on the banks of the Byſtrikz. An anonymous author of a Memoir in the *Journal de Phyſique* for Auguſt 1776, goes farther than Aldrovandus, and, after having compared the penduline and long-tailed titmouſe, obſerves a great analogy between the two birds. Yet had he followed the method of compariſon exactly, he would have perceived that the penduline titmouſe has its bill and legs proportionally longer, its tail ſhorter, its alar extent alſo, and its plumage different.

the

the ſame with what occur in Lithuania, Volhinia, Poland, and Germany. The peaſants regard them with ſuperſtitious veneration: one of theſe neſts is ſuſpended near the door of each cottage, and the poſſeſſors hold it as a protector from thunder, and its little architect as a ſacred bird. We might almoſt regret that nature is not more ſparing of her wonders: for every extraordinary appearance is a ſource of new errors.

Theſe titmice inhabit likewiſe Bohemia, Sileſia, the Ukraine, Ruſſia, Siberia, and whereever, in ſhort, thoſe plants grow that furniſh the cotton for conſtructing their neſts *. But they are rare in Siberia, according to Gmelin †: nor can they be very common near Bologna; ſince, as we have already obſerved, Aldrovandus was unacquainted with them. Daniel Titius regards Italy ‡ as the original country of the penduline titmice, whence they paſſed into the ſtate of Venice, into Carinthia, Auſtria, the kingdom of Bohemia, Hungary, Poland,

* Daniel Titius remarks that Volhinia, Poleſia, Lithuania, and other cantons of Poland, abound with marſhes, and aquatic plants or trees, ſuch as willows, alders, poplars, knapweeds, ſtarworts, hawkweeds, juncagos, &c.

† *Voyage en Sibérie*, t. ii. p. 203. The Counſellor J. Ph. of Strahlenberg had obſerved theſe birds in Siberia before Gmelin, according to Daniel Titius.

‡ Hence the name *Romiſch*.

and the more northern countries. They almoſt always haunt the marſhes, and hide themſelves among the bulruſhes, and the foliage of the trees which grow in ſuch ſituations. It is aſſerted that they never migrate on the approach of winter *. Such may be the caſe in temperate countries, where inſects are found through the whole year: but, in the northern climates, I ſhould ſuppoſe that the penduline titmice at leaſt change their haunts during the intenſe colds; and, like the other titmice, reſort to the inhabited ſpots. Accordingly, Kramer informs us that, in the vicinity of the city of Pruck, they are much more numerous in winter than in any other ſeaſon; and that they always lodge, from preference, in the bulruſhes and reeds.

It is ſaid that they have a warble, though not well known. Yet the young ones taken from the neſt have been kept ſeveral years, and fed ſolely on ants' eggs †: therefore they ſing not in the cage.

The plumage of this bird is very ordinary: the crown of the head is whitiſh; the back of

* Cajetan Monti and Daniel Titius.

† Titius, p. 43, 44. In another place he ſays, that they ſing better than the long-tailed titmouſe, which, according to Belon, has a pleaſant ſong.

the

the head, and the upper ſide of the neck, cinereous *; all the upper ſide of the body gray, but tinged with ruſty in the fore part; the throat, all the under ſide of the body, white, tinged with cinereous gray before, and ruſty behind: there is a black bar on the face, that extends on both ſides to the eyes, and much beyond them; the ſuperior coverts of the wings brown, edged with rufous, which becomes gradually more dilute near the extremity; the quills of the tail and of the wings brown alſo, but edged with whitiſh; the bill cinereous; the legs reddiſh cinereous.

It appears, from the deſcription of Cajetan Monti, that in Italy the penduline titmice have more rufous in their plumage, and a ſlight tinge of green on the ſuperior coverts of their wings, &c. and, from that of Gmelin, that thoſe in Siberia have the back brown, the head white, and the breaſt tinged with rufous. But theſe are only variations occaſioned by the climate, or perhaps owing to the difference of deſcription; for a conſiderable diverſity of appearance will ariſe from the diſtance they are held from the eye, and the light with which they are viewed.

* Daniel Titius ſaw a blackiſh ſpot near the firſt vertebra of the neck, and another near the anus.

The female, according to Kramer, is without the black bar as in the male: but, according to Gmelin, it has that bar. In both, the iris is yellow, and the pupil black: and they are ſcarcely larger than the common wren; that is, they are nearly the ſame ſize with the blue titmouſe.

Total length, four inches and a half; the bill, five lines; the upper mandible a little incurvated; the lower longeſt in the young ones*; the *tarſus,* ſix lines and an half; the nails very ſharp, the hind one longeſt; the alar extent, ſeven inches and one third; the tail two inches, conſiſting of twelve quills, a little tapered, and exceeding the wings thirteen lines. [A]

[A] Specific character of the Remiz, *Parus Pendulinus:* "Its head is ſomewhat ferruginous, with a black ſtripe on the eyes; the quills of the wings and of the tail brown, either margin ferruginous."

* Titius, p. 19 and 23.

THE

LANGUEDOC TITMOUSE.

La Penduline, *Buff.* *
Parus Narbonenſis, *Gmel.*

MONTI ſuppoſed that the Remiz, or penduline titmouſe, was the only European bird that ſuſpended its neſt from branches †: but, not to mention the golden oriole, which faſtens its neſt ſometimes to ſlender boughs, and which Friſch has miſtaken for the long-tailed titmouſe ‡, there is another ſpecies well known in Languedoc, though entirely overlooked by the naturaliſts, which builds as artfully as the Poliſh titmouſe, and diſplays even more ſagacity in the ſtructure. It deſerves the more our attention, becauſe its talents have not attained the ſame celebrity. It may be conſidered as analogous

* In Languedoc it is vulgarly called *the wild canary* The bird and neſt were ſent by M. de Brouſſe, Mayor c Aramont, Deputy of the States of Languedoc.

† See Collection Academique partie etrangere, t. x. p 371. *Academie de Bologne.*

‡ This error was the eaſier avoided, ſince the neſt of the oriole is cup-ſhaped, open above; and ſince that bird never employs the down furniſhed by the flowers and leaves or certain plants, even though they grow abundantly near it.

logous to the penduline titmouſe, but not as a mere variety: the differences in its ſize, its proportions, its colours, the ſhape of its neſt, &c. are fully ſufficient to conſtitute it another ſpecies.

The neſt is very large, compared with the ſize of the bird; ſhut above, and has nearly the bulk and ſhape of an oſtrich's egg: its greater axis is ſix inches; its ſmaller axis, three and an half. It is ſuſpended from the fork of a flexible poplar branch; and, to give greater ſtability, it is wrapped with wool, for the ſpace of more than ſeven or eight inches; and the down of the poplar, the willow, &c. is alſo uſed. The aperture is placed in the ſide near the upper part, and has alſo a ſort of projection, or penthouſe, that juts eighteen lines. From theſe precautions, the young are better ſheltered and concealed, and conſequently ſafer than thoſe of the penduline titmouſe.

The throat and all the under ſide of the body are ruſty white; the upper ſide ruſty gray, deeper than the under; the top of the head gray; the ſuperior coverts of the wings blackiſh, edged with rufous, and likewiſe the middle quills: but the rufous grows more dilute near the end. The great quills blackiſh, edged with whitiſh; the quills of the tail blackiſh, edged with

with light rufous: the bill black; the upper ridge brownifh yellow; the legs lead colour.

Total length, rather lefs than four inches; the bill, like that of the titmoufe, and four lines and an half; the *tarfus*, fix lines; the hind nail the ftrongeft of all, a little hooked: the tail, eleven or twelve lines, would be exactly fquare, were not the two exterior quills a little fhorter than the reft: it exceeds the wings fix lines.

THE

LONG-TAILED TITMOUSE.

La Meſange à longue Queue, *Buff.*
Parus Caudatus, *Linn. Gmel. Kram. Friſ. Ray, Klein & Scop.*
Parus Longicaudus, *Briſſ.*
Schwanzmeiſe, Pfannenſtiel, *Gunth. & Wirs.*
Pendolino, o Paronzino, *Zinn.**

THIS very ſmall bird is beſt diſtinguiſhed by the length of its tail, which exceeds that of its body. It is ſlender, and its flight ſo rapid, that it ſeems like a dart ſhooting through the air†. But notwithſtanding this remarkable diſparity ‡, it has ſtill the eſſential characters of

* In Greek, Αιγιθαλος ορεινος, or mountain titmouſe; *Ariſt.* Hiſt. Anim. lib. viii. 3. In Italian, *Paronzino, Pulzonzino, Pendolino* in German, *Schwantz-maiſe*, (tail-titmouſe), *Zagel-maiſe*, (the ſame), *Pfannen-ſtiel*, or *Pfannen-ſtiegliz* (tail-ſiſkin), *Riet-maiſe* (reed-titmouſe), *Berg-maiſe* (mountain-titmouſe), *Schnee-maiſe* (ſnow-titmouſe): in Poliſh, *Sikora Zdtugim Ogonem.*

† Britiſh Zoology.

‡ This diſparity induced Ray to ſuppoſe it a diſtinct genus. The authors of the Britiſh Zoology obſerve, that it much reſembles the ſhrikes in the form of its bill, which is more convex than that of the titmice, and by other minute analo-

N.° 133

THE LONG-TAILED TITMOUSE.

of the titmice: its bill is ſhort, yet pretty ſtrong: its chief reſidence is in the woods; it is active and reſtleſs, fluttering inceſſantly from buſh to buſh, from ſhrub to ſhrub; running among the branches, hanging by the feet; lives in ſociety, attending to the call of its companions; feeds on caterpillars, flies, and other inſects, ſometimes ſeeds; crops the buds from the trees; lays a great many eggs: in fine, according to the moſt accurate obſervations, it has the principal exterior characters of the titmice; and, what is more deciſive, it has their habits and œconomy. Nor is the long tapered tail entirely diſtinct from thoſe of all the other titmice, ſince thoſe of the bearded and penduline kinds are ſomewhat of the ſame form.

With regard to the mode of conſtructing its neſt, it follows a middle plan between that of the ox-eye and colemouſe and that of the penduline titmouſe. It does not conceal it in the hole of a tree, which would be ill adapted for its long tail, and it ſeldom or never hangs it to a ſlender ſtring; but it fixes it firmly in

analogies: but there is evidently a wide difference between a ſhrike and the long-tailed titmouſe. I am ſurpriſed that ſome nomenclator has not ranged the latter among the *motacillæ*, where it would have made a figure, its long tail vibrating briſkly and repeatedly up and down.

the branches of ſhrubs, three or four feet from the ground; works it into an oval, and almoſt cylindrical form; cloſes it above, leaving only an aperture of an inch in the ſide, and often makes two oppoſite holes, to avoid the inconvenience of turning *; a precaution the more uſeful, as the tail quills are eaſily detached, and drop with the ſlighteſt ruffling †. The neſt differs from that of the penduline titmouſe in other circumſtances alſo: it is larger ‡, of a more cylindrical form; its texture is not ſo cloſe; its little aperture has ſeldom the protuberant brim §; its outer coat conſiſts of ſtalks of herbs, moſs, lichens, and, in ſhort, of the coarſer materials; and the inſide is lined with a great deal of feathers, and not with that cotton which the willows and other plants furniſh the penduline titmouſe.

The long-tail titmouſe lays ten or fourteen eggs, and even as many as twenty, concealed almoſt entirely beneath the feathers collected in the bottom of the neſt; theſe eggs are of the ſize of a hazel nut, their greateſt diameter being ſix lines; they are ſurrounded by a reddiſh zone

* Friſch and Rzaczynſki.

† Hence it has been called *loſe-tail* (perd-ſa-queue).

‡ I meaſured one of theſe neſts, which was eight inches high, and four wide.

§ Cajetan Monti and Daniel Titius, compared together.

on

on a gray ground, which grows more dilute at the thick end.

The young continue with the parents through the whole winter, and hence the ſmall flocks of twelve or fifteen ſeen during that ſeaſon. They utter a ſmall ſhrill cry, only as a call; but in the ſpring they acquire a new modulation, which makes it much more pleaſant*.

Ariſtotle aſſures us, that theſe birds prefer the mountains. Belon ſays, that he obſerved them in all countries; and Belon had travelled: he adds, that they ſeldom leave the woods to viſit the gardens. Willughby informs us, that in England they frequent the gardens more than the mountains: Hebert aſſerts the ſame, but reſtricts it to the winter ſeaſon. According to Geſner, they appear during cold weather only, and haunt the marſhes; whence their name *reed titmice*. Daubenton, the younger, has ſeen flights of them in the king's garden, about the end of December; and has informed

* "It ſings ſo pleaſantly in the ſpring," ſays Belon, "that ſcarce any bird has a voice ſo lofty and airy." Geſner aſſerts that, in this ſeaſon, the long-tailed titmouſe utters *guickeg, guickeg:* this ſeems not to be the *pleaſant ſong* which Belon mentions. Others ſay that this titmouſe has a feeble voice, and a little ſhrill cry, *ti, ti, ti, ti;* but this is certainly not the warble heard in ſpring.

me, that they are frequent in the woods of Boulogne. Laftly, fome maintain that they continue through the winter; others, that they migrate; and others, that they arrive later than the other titmice, and have therefore been called *fnow titmice*. Thefe oppofite ftatements may be reconciled, by fuppofing, what is at the fame time very probable, that thefe birds vary their range according to circumftances; that they remain when their fituation is comfortable, and change when they want a better; that they inhabit the mountain or the vale, the bank or the marfh, the foreft or the vineyard, or wherever they can enjoy convenience and fubfiftence. They are feldom caught in traps, and their flefh is unpleafant food.

Their feathers are loofe, and refemble a very long down: they have a fort of black eyebrows, and the upper eyelids of an orange yellow, but this colour is hardly vifible in dried fubjects; the upper fide of the head, the throat, and all the under fide of the body, white, fhaded with blackifh on the breaft, and fometimes tinged with red on the belly, the loins, and under the tail: the back of the neck is black, whence rifes a bar of the fame colour, which ftretches through the whole of the upper part of the body between two broad bars of baftard red; the

the tail black, edged with white; the fore part of the wing black and white; the great quills blackiſh; ſo are the middle ones, but edged with white, except thoſe next the body, which are of the ſame rufous with the black; the ground of the feathers deep cinereous; the iris gray; the bill black, but gray at the point; and the legs blackiſh.

The white bar on the crown of the head ſpreads more or leſs, and ſometimes incroaches ſo much on the lateral black bars, that the head appears entirely white. In ſome ſubjects, the under ſide of the body is all white; ſuch were thoſe ſeen by Belon, and ſome that I have obſerved myſelf. In females, the lateral bars of the head are only blackiſh, or even variegated with black and white, and the colours of the plumage are not well defined or contraſted. The bird is hardly larger than the gold-creſted wren; it weighs about one hundred and fourteen grains. As its feathers are almoſt always briſtled, it appears rather thicker than in reality.

Total length, five inches and two-thirds; the bill, three lines and a half, thicker than that of the blue titmouſe, the upper mandible a little hooked; the tongue rather broader than in that bird, and terminated with filaments; the *ta ſus*, ſeven lines and a half; the hind nail the

 ſtrong-

ſtrongeſt; the alar extent, ſix inches and a half; the tail, three lines and a half, conſiſting of twelve unequal quills, and irregularly tapered, increaſing conſtantly in length from the outermoſt which is eighteen lines to the fifth which is forty-two, or thereabouts; the intermediate pair only thirty-nine at moſt, and hardly equal to the fourth: the tail exceeds the wings about two inches and a half.

Its inteſtinal tube, four inches; only a ſlight trace of a *cæcum*; the gizzard muſcular, and contained portions of inſects and vegetables, a bit of a nut, but no pebbles. [A]

[A] Specific character of the Long-tailed Titmouſe, *Parus Caudatus*: "Its top is hoary, its tail longer than its "body, its head bearded."

THE

CAPE TITMOUSE.

Le Petit Deuil, *Buff.*
Parus Capenſis, *Gmel.*

THIS little titmouſe was brought from the Cape of Good Hope by Sonnerat, who publiſhed a deſcription of it in the *Journal de Phyſique.* Its plumage is black, gray, and white; the head, the neck, the upper and under ſides of the body, light cinereous gray; the quills of the wings black, edged with white; the tail black above, and white below; the iris, the bill, and the legs black.

This bird reſembles thoſe preceding, particularly the long-tailed titmouſe, in the manner it conſtructs its neſt; which it places in the thickeſt buſhes, but not at the extremity of the branches, as ſome naturaliſts have ſuppoſed. The cock aſſiſts the hen in building; he ſtrikes his wings forcibly againſt the ſides of the neſt, and compacts it into the form of an elongated ball; the aperture is in the ſide; the eggs in the centre, where they are ſafeſt and warmeſt.

So far the conſtruction agrees with the long-tailed titmouſe: but there is beſides a ſmall compartment where the male lodges during the incubation. [A]

[A] Specific character of the *Parus Capenſis*: "It is "of a dilute cinereous gray; its wing quills black, edged "with white; its tail black above, and white below."

THE

THE

SIBERIAN TITMOUSE.

La Meſange à Ceinture Blanche, *Buff.*
Parus Sibericus, *Gmel.*

WE ſaw this bird in M. Mauduit's cabinet; but we are unacquainted with its hiſtory. Muller makes no mention of it, and perhaps it is not found in Denmark, though it was ſent from Siberia. On the throat and the fore part of the neck is a black mark, which deſcends on the breaſt, accompanied on each ſide with a white bar, which riſes from the corners of the mouth, paſſes below the eye, and deſcends ſpreading as far as the wings, and extends, on each ſide, on the breaſt, where it aſſumes a cinereous hue, and forms a broad cincture: all the reſt of the under ſide of the body is ruſty gray; ſo is the upper ſide, but deeper; the upper part of the head and neck, brown gray; the ſuperior coverts of the wings, their quills, and thoſe of the tail, aſh brown: the quills of the wings, and the outer quill of the tail, edged with rufous gray; the bill and legs, blackiſh.

Total

Total length, five inches; bill, ſix lines; the *tarſus*, ſeven lines; the tail twenty-two lines, and exceeds the wings fifteen lines: it is a little tapered, in which reſpect this ſpecies reſembles more the bearded, the penduline, and the long-tailed titmice, than the others, in all which it is a little forked.

THE

CRESTED TITMOUSE*.

La Mesange Huppée, *Buff.*
Parus Cristatus, *All the Naturalists.*
The Juniper Titmouse, *Charleton.*

IT has a handsome black and white crest, which rises eight or ten lines on the head, and whose feathers taper with an elegant regularity. The bird is also naturally perfumed, and exhales the scents of the junipers and other resinous trees and shrubs, among which it almost constantly lodges†. And these advantages, which seem appropriated to the luxury of society, are enjoyed in the wildest solitude; not so completely, perhaps, but surely in more tranquillity. Forests and heaths, especially those abounding with junipers and firs, are its fa-

* In German, *Schopf-meise* (tufted titmouse), *Hauben-meise* (capped titmouse), *Heiden-meislin* (heath titmouse), *Struss-meislin* (strutting titmouse), *Kopf-meise* (head titmouse): in Swedish, *Tofs-mussa, Tofs-tita, Meshatt*: in Polish, *Sikora-czubata.*

† Charleton.

vourite

vourite haunts: there it lives ſequeſtered, and ſhuns the company of all other birds, even thoſe of its own ſpecies *. It is equally beyond the reach of man: its retreat and its caution ſave it from the fowler's ſnares. It is ſeldom caught in traps; and, if ſurpriſed, it will refuſe food, and, ſpurning every ſoothing attention, will expire in confinement. Hence it is little known: we learn only that, in its congenial ſolitude, it feeds upon the inſects which it catches as they fly about the trees; and that it has the chief character of the titmice, great fecundity.

Of all the French provinces, Normandy is that where it is moſt common. It is unknown, ſays Salerne, both in Orleanois, and in the neighbourhood of Paris. Belon makes no mention of it, nor Olina; and Aldrovandus would ſeem to have never ſeen it: ſo that Sweden, on the one ſide, and the north of France on the other, are the limits of its excurſions.

* This is the opinion of Friſch, corroborated by that of the Viſcount de Querhoënt. Yet I muſt own, that, according to Rzaczynſki, the creſted titmice go in flocks; but his authority will not outweigh that of the other two obſervers. Rzaczynſki adds, that, in autumn, many of theſe birds are caught in the mountains.

Its

Its throat is black, its face white, and alſo its cheeks, on which the white is framed in a ſlender black collar, which riſes from the two ſides of the black mark on the throat, and mounts in a curve to the back of the head. there is a black vertical bar behind the eye; the under ſide of the body, whitiſh; the flanks, light rufous; the upper ſide of the body, rufous gray; the ground of the feathers, black; the quills of the tail gray, and thoſe of the wings brown; all of them edged with rufous gray, except the great ones of the quills, which are partly dirty white; the bill blackiſh, and the legs lead coloured.

Willughby perceived a greeniſh tinge on the back, and on the outer edge of the quills of the tail and of the wings: Charleton ſaw a ſimilar tinge on the feathers that compoſe the creſt; probably theſe feathers have different reflections, or vary ſlightly from age, ſex, &c.

This bird weighed about the third of an ounce, and was hardly larger than the long-tailed titmouſe.

Total length, four inches and three quarters; the bill, five lines and a half; the tongue terminated by four filaments; the *tarſus*, eight lines; the hind nail the ſtrongeſt; the alar

 extent,

extent, ſeven inches and a half; the wing compoſed of eighteen quills; the tail about twenty-two lines, a little forked, and compoſed of twelve quills; it exceeds the wings ten lines. [A]

[A] Specific character of the Creſted Titmouſe, *Parus Criſtatus*: "Its head is creſted; its collar black; its belly white." Its egg is whitiſh rufous, with ſmall reddiſh ſpots.

FOREIGN

FOREIGN BIRDS

WHICH ARE RELATED TO THE TITMICE.

I.

THE CRESTED TITMOUSE OF CAROLINA.

La Mesange Huppée de la Caroline, *Buff.*
Parus Bicolor, *Linn.* & *Gmel.*
Parus Carolinensis Cristatus, *Briss.*
The Crested Titmouse, *Catesby.*
The Toupet Titmouse, *Penn.* & *Lath.*

THE crest of this foreign bird is not permanent, but only rises and tapers to a point during a fit of passion; ordinarily its feathers recline flat on the head.

This bird inhabits, breeds, and continues during the whole year in Carolina, Virginia; and probably it occurs also in Greenland *, since Muller has inserted it in his Danish Zoology. It lives in the forests, and, like all the other titmice, it feeds on insects: it is larger than the preceding species, and differently pro-

* The Greenlanders call it *Auingursak.*

portioned;

portioned; for its bill is ſhorter, and its tail longer: it weighs about four gros: its plumage is pretty uniform: its forehead is encircled with a ſort of black fillet: the reſt of the upper ſide of the head and of the body, and even the quills of the tail and of the wings, are deep gray: the under ſide of the body is white, mixed with a ſlight tinge of red, which becomes more perceptible on the inferior coverts of the wings: the bill is black, and the legs lead coloured.

The female is exactly like the male.

Total length, about ſix inches; the bill, five lines and an half; the *tarſus*, eight lines and an half; the hind nail ſtrongeſt; the tail two inches and an half, conſiſting of twelve quills; and it exceeds the wings about ſixteen lines. [A]

[A] Specific character of the *Parus Bicolor*: "Its head is creſted, and black before; its body cinereous; below tawny whitiſh." It inhabits the foreſts in Virginia and the Carolinas: its flight is ſwift; its note feeble.

II. THE

II.

THE COLLARED TITMOUSE.

La Mefange à Collier, *Buff.*
Parus Carolinenfis Torquatus, *Briff.*
The Hooded Titmoufe, *Catefby.*

IT appears to have a black hood fet a little back on its yellow head, the fore part being uncovered: the throat too has a yellow mark, below which is a black collar: all the reft of the under fide of the body is alfo yellow, and all the upper fide olive: the bill is black, and the legs brown. The bird is nearly the fize of the chaffinch. It inhabits Carolina.

Total length, five inches; the bill, fix lines; the *tarfus*, nine lines; the tail twenty-one lines, a little forked, and exceeds the wings ten lines.

III.

THE YELLOW RUMP TITMOUSE.

La Mesange à Croupion Jaune, *Buff.*
Parus Virginianus, *Linn. Gmel. Briss.*
Luscinia Uropygio Luteo, *Klein.*
The Yellow Rump, *Catesby.*
The Virginian Titmouse, *Penn. & Lath.*

IT creeps on the trees like the woodpeckers, says Catesby; and, like them, it commonly feeds upon insects: its bill is blackish; its legs brown; its throat, and all the under side of its body, gray; its head, and all the upper side of its body, as far as the end of the tail, including the wings and their coverts, of a greenish brown, excepting always the rump, which is yellow. This yellow rump is the only beauty of the bird, and what alone interrupts the insipid uniformity of its plumage. The female resembles the male: both are nearly as large as the chaffinch, and were observed in Virginia by Catesby.

Total length, about five inches; the bill, five lines; the *tarsus*, eight lines; the tail twenty-one lines, a little forked, consisting of twelve quills, of which the intermediate ones are a

little

little ſhorter than the lateral ones, and it exceeds the wings about ten lines. [A]

[A] Specific character of the *Parus Virginianus*: "Its rump is yellow; its body cinereous."

IV.

THE YELLOW-THROATED GRAY TITMOUSE.

La Meſange Griſe à Gorge Jaune, *Buff.*
Parus Carolinenſis Griſeus, *Briſſ.*
The Yellow-throated Creeper, *Cateſby.*

NOT only the throat, but all the fore part of the neck, is of a fine yellow; and alſo on each ſide of the head, or rather at the baſe of the upper mandible, is a ſmall trace of that colour: the reſt of the under ſide of the body is white, with ſome black ſtreaks on the flanks: all the upper ſide is of a handſome gray: a black bar covers the face, extends on the eyes, and deſcends on both ſides of the neck, accompanying the yellow mark of which I have ſpoken: the wings are of a brown gray, and bear two

 white

white ſpots: the tail is black and white; the bill black, and the legs brown.

The female has neither the fine yellow which ſets off the plumage of the male, nor the black ſpots which riſe out of the other colours.

This bird is common in Carolina. It weighs only two gros and an half; and yet Briſſon ſuppoſes it as large as the ox-eye, which is ſeven or eight gros.

Total length, five inches and one third; the bill, ſix lines; the *tarſus*, eight lines and an half; the nails very long, the hind one ſtrongeſt; the tail twenty-ſix lines, a little forked, conſiſting of twelve quills, and exceeds the wings fourteen lines.

V.

THE GREAT BLUE TITMOUSE.

La Groſſe Meſange Bleue, *Buff.*
Parus-Cyanus, *Gmel.*
Parus Indicus, *Ray & Will.*
Parus Cæruleus Major, *Briſſ.*

THE figure of this bird was communicated by the Marquis Fachinetto to Aldrovandus. It made a part of the coloured drawings of of birds, which certain travellers from Japan presented to Pope Benedict XIV. and which were suspected by the sagacious naturalist Willughby, as either imaginary or very inaccurate. But we shall exactly copy the description of Aldrovandus.

Light blue predominates on all the upper part of this bird, and white on the under: a very deep blue is spread on the quills of the tail, and of the wings: the iris is yellow: there is a black spot behind the eyes: the tail is as long as the body, and the legs are black and small, which is not the case in the titmouse: besides, the description shews a certain uniformity different from the design of nature, and which justifies the suspicions of Willughby. [A]

[A] A bird, answering nearly to Aldrovandus's description, has lately been discovered in the remote parts of Russia, and described, in the Petersburg Transactions, by Pallas, Falck, and Lepechin. We cannot do better than transcribe Mr. Pennant's account of it.

"AZURE TITMOUSE. With a very short and thick bill; crown, and hind part of the head, of a hoary whiteness; the lower part of the last bounded by a transverse band of dark blue; cheeks white, crossed by a deep blue line, extending beyond the eyes; back, light blue; rump, whitish; under side of the neck, breast, and belly of a snowy whiteness, with a single dusky spot on the breast: wings varied with rich blue, dusky, and white; tail rather long, of a dusky hue, tipped with white; legs dusky blue."

"Size of the English Blue Titmouse. The plumage of

 this

this elegant ſpecies is extremely looſe, ſoft, and of moſt exquiſitely fine texture; and ſo liable to be raiſed, that when the bird is ſitting, but eſpecially when it is aſleep, it appears like a ball of feathers."——

"It inhabits, in great abundance, the northern woods of Siberia and Ruſſia, and about Synbirſk, in the government of Kaſan. It is a migratory bird, and appears in winter converſant about the houſes in Peterſburgh: it twitters like the common ſparrow, but with a ſofter and ſweeter note." *Arctic Zoology*, vol. ii. 426.

VI.

THE AMOROUS TITMOUSE*.

La Meſange Amoureuſe, *Buff.*
Parus Amatorius, *Gmel.*
Parus Eraſtes, *Commerſon.*

CHINA alſo has its titmice: the preſent was brought by the Abbé Gallois from the extremity of Aſia, and was ſhewn to Commerſon in 1769. I have been induced, by the account of this naturaliſt, to place it at the cloſe of the liſt of titmice, from which it manifeſtly differs, by the length and ſhape of its bill.

The epithet of amorous expreſſes the warmth

* Some have called it the *Cancneſs*, on account of its garb.

of

of its conſtitution: the male and female continually careſs each other, at leaſt in the cage, where it is their ſole occupation: they even wear out their vigour; and if this ſolacement baniſhes the gloom of confinement, it alſo abridges the period of their life. Commerſon does not tell us if the ſame ardour pervades their other functions, and is diſplayed in the conſtruction of their neſt, their incubation, &c. and whether their brood is as numerous as in the other titmice. In the ordinary courſe of nature the affirmative is the moſt probable, though it may indeed be modified by climate, and peculiar inſtincts.

Their plumage is entirely of a ſlaty black, which appears equally on the upper and under ſide of the body, and the uniformity of which is only interrupted by a bar divided by yellow and rufous, placed longitudinally on the wing, and formed by the exterior border of ſome of the middle quills: this bar has three indentings, at its origin, near the middle of the wing, which conſiſts of fifteen or ſixteen quills differing little in length.

The amorous titmouſe weighs three gros: it is of the ſame ſhape with the other titmice, and of a middle ſize; but its tail is ſhorter, and only five inches and a quarter; the bill eight lines, black at the baſe, and bright orange at the oppoſite extremity; the upper mandible exceeding a little the lower, and having its edges ſlightly indented near the tip, the tongue ſomewhat truncated at the end, as in the other titmice; the *tarſus*, eight lines; the mid toe the longeſt, adhering by its firſt phalanx to the outer toe; the nails forming a ſemi-circle by their curvature, the hind one ſtrongeſt; the alar extent, ſeven inches and an half; the tail near two inches, a little forked, conſiſting of twelve quills: it exceeds the wings more than an inch. [A]

[A] Specific character of the *Parus Amatorius*: "It is tinged with ſlate colour; a longitudinal ſpot on the wings, divided by yellow and rufous."

THE BLACK TITMOUSE, or the *Parus Cela* of Linnæus *, bears a ſtriking reſemblance to this ſpecies; ſince the only difference in regard to colours is, that its bill is white, and there is a yellow ſpot on the ſuperior coverts of the tail. Linnæus ſays that it is found in the Indies; but he means the Weſt Indies, for Dupratz ſaw it in Guiana. Notwithſtanding this, the wide

* The Guiana Titmouſe of Latham. Specific character "It is black; its bill white; a yellow ſpot on the wings, and at the baſe of the tail."

difference

difference of climate, we muſt conſider it as only a variety of the amorous titmouſe of China; but to be more deciſive would require a knowledge of its ſize, proportions, and, above all, of its natural habits.

THE

NUTHATCH.

La Sitelle, *Buff.*
Sitta Europæa, *Linn. & Gmel.*
Parus Facie Pici, *Klein*
Picus Subcæruleus, *Schwenckf.*
Picus Cinereus, *Gesner & Johnst.*
Sitta, *Ray, Charl. Sibb. Fris. Briss. &c.*
The Woodcracker, *Plot.*
The Nutbreaker, *Albin.* *

THE nuthatch seldom migrates from one country to another; it constantly resides where it is bred, only in winter it seeks warm aspects, approaches the dwellings, and even visits sometimes the vineyards and the gardens. Perhaps it finds shelter in the same holes where it collects its stores, and probably passes the night; for, when kept in confinement,

* In Greek, Σιττη or Σιπτη; Arist. *Hist. Anim.* lib. ix. 1 & 17. Also Υλοτομουσα, from ὑλη, wood, and τεμνω, to cut; Κιναιδος, from κινεω, to move; and Σεισοπύγης, from σειω, to shake, and πυγη, the rump, on account of a motion which it has with its tail: in modern Greek, Καρκουιστης: in Latin, *Sitta*: in Italian, *Pico*, or *Picchio*, *Ziollo*: in German, *Nuss-backer*, *Nusshaer*, *Nussbicker*, *Nussbickel* (these words signify *nutcracker*), *Meyspecht* (May-spight), *Blaw-spechtle* (blue spight-

No. 134

THE EUROPEAN NUTHATCH.

ment, though it ſometimes perches on the bars of the cage, it ſeeks holes to ſleep in, and, if unſucceſsful, will repoſe even in the drawer. It ſeldom ſquats in the natural poſition, or with its head erect; it is commonly awry, or even hanging downwards; in this ſituation it bores the nuts, after fixing them firm in a chink*. It trips on the trees in all directions to catch inſects. Ariſtotle ſays, that it habitually breaks the eagle's eggs; and indeed, if it could climb to the lofty eyries of the king of birds, it might pierce and eat the eggs, which are not ſo

ſpightling), *Groſſe Baum-Kletter* (great wood-climber): in Swediſh, *Noetwaecka*, *Noetpacka*: in Daniſh, *Spætmeiſſe*: in Norwegian, *Nat wacke*, *Edge*, *Eremit*: in Poliſh, *Dzieciot Modrawy*.

Moſt of the names which the moderns have impoſed on this bird, convey falſe or incomplete ideas: ſuch are the *May Woodpecker*, *Blue Woodpecker*, *Maſon Woodpecker*, *Nutcracker*, *Creeper*, *Wagtail*, &c. This bird ſtrikes with its bill on the bark of trees with more force and noiſe than even the woodpeckers and titmice, and it has much of the air and aſpect of the latter: but it is diſtinguiſhed from the titmice by the form of its bill, and from the woodpeckers by the form of its tail, its feet, and its tongue. It is diſcriminated from the creepers by its bill and its habit of cracking nuts; and from the nutcrackers again by its habit of creeping upon trees: it has an alternate motion of the tail upwards and downwards, like the wagtails, but its œconomy is entirely different.

* Albin.

hard as nuts. But it is idle to aſſert, that this is what provokes the vengeance of the eagle*, as if a bird of prey needs inſtigation to plunder and devour weaker birds.

Though the nuthatch ſpends a great part of its time in climbing or creeping upon trees, its motions are quicker and nimbler than thoſe of the ſparrow; they are alſo ſmoother and more connected, for it makes leſs noiſe in flying. It commonly lives in the woods, in the moſt ſolitary manner; yet if confined in a volery with other birds, ſparrows, for inſtance, or finches, it will continue on good terms with them.

In ſpring, the male has a ſong, or love-call, *guiric, guiric*, often repeated. The female is long invited, it is ſaid, before ſhe will yield to the ſolicitations; but, when the union is accompliſhed, both in concert labour in forming their neſt. They fix it in the hole of a tree†; and when they cannot find one that ſuits them, they make an excavation with their bills, if the wood be worm-eaten: if the external aperture

* *Hiſt. Anim.* lib. ix. 1, Perhaps the *clamitoria* or *prohibitoria* of Pliny is the ſame bird: the name *prohibitoria* might allude to the ancient fables, with regard to the nuthatch, and to its uſe in necromancy. See *Not Hiſt.* lib. x. 14.

† Sometimes in the hole of a wall or of a roof, ſays Linnæus.

be

be too large, they contract it with fat earth, and ſometimes with dirt, which they mould and faſhion, it is ſaid, as a potter would do his clay, and they ſtrengthen it with ſmall ſtones; and hence are derived the names *maſon-woodpecker* and *wipe-pot* *. From the appearance of the outſide of the neſt, we could hardly ſuppoſe it lodged birds.

The hen lays five, ſix, or ſeven eggs of the ordinary ſhape, of a dirty white ground, and dotted with ruſty; the bed is wood duſt, moſs, &c. She hatches aſſiduouſly; and ſo ardent is her attachment that ſhe will ſuffer the feathers to be torn, rather than quit the neſt. If it be rummaged with a ſtick, ſhe briſtles with rage, and hiſſes like a ſerpent, or rather like a tit-mouſe in the ſame ſituation. Nor does ſhe leave her eggs to ſeek food, but patiently expects the return of the male, who affectionately brings ſupply. They do not live altogether on ants, like the woodpeckers, but on caterpillars, beetles, gadflies, and all ſorts of inſects, and the various kinds of nuts†. The fleſh of the young ones

* *Torche-pot*: this word comes from *torche-poteux*, which ſignifies *wipe-hole*.

† I fed a female ſix weeks with hemp-ſeed which other birds had dropped bruiſed. It has indeed been remarked, that the nuthatch viſits the hemp-fields about the month of September.

accord-

accordingly, when they are fat, is excellent, and has not the wild taſte of the woodpeckers.

The incubation is completed in May*; and after the young are reared, the parents ſeldom begin a ſecond hatch, but diſſolve their union, and live ſeparate during the winter. "The peaſants have obſerved," ſays Belon, "that the cock beats the hen whenever he finds her after parting; and hence it is become a proverb, that a perſon who keeps his wife in due order is like a nuthatch." But the conduct of the huſband has no concern, I am confident, in the preſent caſe: the female, as ſhe is the laſt to pair, is probably the firſt in ſeparating; and when the male meets her after a long abſence, he loads her with careſſes, and gives vent to rapturous endearments, which inaccurate obſervers may miſtake for harſh uſage.

The nuthatch is ſilent through the greateſt part of the year; its ordinary cry *ti*, *ti*, *ti*, *ti*, *ti*, *ti*, *ti*, which it repeats as it ſcrambles round the trees, and quickens the meaſure more and more. Linnæus aſſerts from the teſtimony of Strom, that it cries alſo during the night.

Beſides the different cries and the noiſe which it

* I ſaw ſome neſtlings by the 10th, and I ſaw ſome eggs not hatched by the 15th.

it makes in beating on the bark, the nuthatch, inſerting its bill in a chink, makes another very ſingular ſound, as if the tree were ſplit in two, and which may be heard more than two hundred yards*.

This bird has been obſerved to hop; to ſleep with its head under its wing; to ſpend the night on the floor of its cage, though there were two rooſts where it might have perched. It it ſaid not to frequent ſprings, and therefore cannot be enſnared by placing limed twigs near theſe. Schwenckfeld relates that it is often caught by uſing tallow alone for the bait: and this is another feature of reſemblance to the titmice, which, as we have ſeen, are fond of all fat ſubſtances.

The cock weighs near an ounce, and the hen five or ſix gros only †. In the former, all the upper part of the head and body, and even the two intermediate quills of the

* *Britiſh Zoology*. Beſides their *toc, toc, toc*, againſt the wood, theſe birds rub their bill upon the dry and hollow branches, and make a noiſe *grrrrrro*, which is heard at a very great diſtance, and might be imagined to proceed from a bird twenty times as large: this I have been aſſured by an old gamekeeper, who moſt aſſuredly had never read the Britiſh Zoology.

† One dried in the chimney a year, and very well preſerved, weighed only two gros and a half.

tail

tail, are blueiſh cinereous; the throat and the cheeks whitiſh; the breaſt and the belly orange; the flanks, the thighs, and near the anus, of a deeper caſt, inclined to cheſnut; the lower coverts of the tail whitiſh, edged with rufous, and extending five lines from the end of the tail: there is a black bar which riſes from the noſtrils, paſſes over the eyes, and extends behind the ears; the great ſuperior coverts, and the quills of the wings brown, edged with gray, which is more or leſs intenſe; the lateral quills of the tail black terminated with cinereous, the outermoſt edged with white as far as the middle, and croſſed near the end with a ſpot of the ſame colour; the three following marked with a white ſpot on the inſide; the bill cinereous above, lighter below; the legs gray; the ground of the feathers blackiſh cinereous.

In the female the colours are fainter. I obſerved one on the third of May, of which all the under ſide of the body, from the anus to the baſe of the neck, was unfeathered, as common in hen birds.

Total length, ſix inches; the bill, ten lines, ſtraight, a little inflated above and below; the two mandibles nearly equal, and the upper one not ſcalloped; the noſtrils almoſt round, half

covered

covered by ſmall feathers, which ſprout at the bottom of the bill, and run parallel to its aperture: the tongue is flat, and broader at the baſe. [A]

[A] Specific character of the Nuthatch or Nutjobber, *Sitta Europæa*: "Its tail quills are black, its four lateral ones white below the tip."

VARIETIES of the NUTHATCH.

THE character of this genus of birds ſeems little affected by the influence of various climates. Its œconomy and natural habits are ever the ſame: the upper part always aſh gray, the lower ſtained with rufous, more or leſs dilute, and ſometimes whitiſh. The chief difference conſiſts in the ſize and proportions, and this does not conſtantly reſult from climate. After attentively comparing the foreign Nuthatches with the European kinds, I am convinced that they are varieties of the ſame ſpecies. I except one only, which differs in many reſpects; and, from the little curvature of its bill, ſeems to form the ſhade between the nuthatches and the creepers.

I. The Little Nuthatch*. I muſt borrow this variety from Belon. According to him, it is much ſmaller than the common nuthatch: its plumage, its bill, and its feet are the ſame. It lives in the woods like the large ſort, and is equally ſolitary. "It is more noiſy, joyous and roving," ſays Belon; "the male is never ſeen with other company but its female; and if it meets with any other of the ſpecies (he means a male), it attacks it, teaſes it, and contends obſtinately till his rival gives way; then cries ſhrill and with all his vigour, inviting his female, and demanding the prize attached to his victory." To ſuch caſes, perhaps, Belon alluded, when he aſſerted that the voice was louder than that of the ordinary nuthatch.

II. The Canada Nuthatch †. It climbs, ſays Briſſon, and runs upon the trees like the European kind, and differs only by the colour of the fillet, which it has whitiſh; but it approaches the other by a blackiſh ſpot behind the eye. On a near examination, we perceive ſome diverſity in the ſhades and propor-

* Sitta Minor, *Briſſ.*

† Sitta Canadenſis, *Linn. Gmel. Briſſ.*

Specific character: "Its eyebrows are white."

tions,

tions, which will beſt be conceived by comparing the figures. It is nearly of the ſame ſize with the preceding variety.

Total length, four inches ſix lines; the bill, ſeven lines and a half; the *tarſus*, ſeven lines; the mid toe, ſix lines and a half, the hind nail the ſtrongeſt; the alar extent, ſeven inches and a quarter; the tail eighteen lines, conſiſting of twelve equal quills, and exceeding the wings eight lines.

III. THE BLACK-CRESTED NUTHATCH*

The black creſt, and a ſort of black and white ſtripes near the end of the quills of the tail, are the principal differences that diſtinguiſh this from the common nuthatch. It has no black fillet; but this is ſuppoſed to be loſt in the edges of a hood of the ſame colour, which covers the head. It inhabits Jamaica, where Sloane obſerved it: it feeds on inſects, like the rock manakin, ſays this philoſophical traveller: it is found among the buſhes in the ſavannas: it is

* Sitta Jamaicenſis, *Linn. Gmel. & Briſſ.*
Sitta Major Capite Nigro, *Ray.*
The Loggerhead, *Brown & Sloane.*
The Jamaica Nuthatch, *Lath.*

Specific character: "It has a black cap."

ſo tame, and allows a perſon to go ſo near to it, that it is often knocked down with ſticks; whence it has been called *the loggerhead*. It is nearly of the ſize of the common nuthatch. Sloane remarks that its head is large.

Total length, five inches five lines; the bill, eleven lines, triangular, compreſſed, ſurrounded at the baſe with little black briſtles; the noſtrils round; the *tarſus* and mid toe, ſeventeen lines, the hind nail the ſtrongeſt; the alar extent, ten inches; the tail, two inches and two-thirds.

IV. THE LITTLE BLACK-CRESTED NUTHATCH*. All that Brown informs us with regard to this bird is, that it inhabits the ſame country with the preceding; that it is ſmaller, but is ſimilar in all other reſpects. Perhaps it was only a young one, not fully grown: the name which Brown applies ſeems to favour this conjecture.

V. THE BLACK-HEADED NUTHATCH† Its habits are the ſame with thoſe of the com-

* Sitta Jamaicenſis, var. 1ſt. *Linn.*
Sitta Jamaicenſis Minor, *Briſſ.*
The Leaſt Loggerhead, *Brown.*

† Sitta Europæa, 2d var. *Linn.*
Sitta Carolinenſis, *Briſſ.*
The Smaller Loggerhead, *Brown.*

mon nuthatch: it clambers both upwards and downwards: it continues the whole year in its native climate, Carolina. It weighs four gros and three quarters: the upper ſide of the head and neck is covered with a ſort of black cowl, and the lateral quills of the tail are variegated with black and white. In other reſpects, its plumage is the ſame with that of the European nuthatch, but has rather more whitiſh beneath the body.

Total length, five inches and a quarter; the bill, nine lines; the *tarſus*, eight lines and a half; the mid toe nine lines, the hind nail the ſtrongeſt; the tail nineteen lines, and does not exceed the wings.

VI. The Little Brown-headed Nuthatch*. I need only add, that there is a whitiſh ſpot behind the head; that the ſuperior coverts of the wings are brown, and that the lateral quills of the tail are of an uniform black. It is alſo much ſmaller than the preceding varieties; and this circumſtance, together with the obvious difference of plumage,

* Sitta Puſilla, *Lath. Ind.*
The Loggerhead, *Sloane.*
The Small Nuthatch, *Cateſby.*

ſufficiently diſtinguiſhes it from Sloane's ſecond ſpecies of nuthatch, though Briſſon ſeems inclined to confound them. It is only two gros: it continues the whole year in Carolina, where it lives on inſects, like the black-headed nuthatch.

Total length, four lines and one-third; the bill, ſeven lines; the tail fourteen lines, conſiſting of twelve equal quills, and hardly exceeds the wings.

FOREIGN

FOREIGN BIRDS

RELATED TO THE NUTHATCH

I.

THE GREAT HOOK-BILLED NUTHATCH.

Sitta Major, *Gmel.*
The Great Nuthatch, *Lath.*

IT is the largeſt of the known nuthatches: its bill, though pretty ſtraight, is inflated at the middle, and a little hooked at the end; the noſtrils are round; the quills of the tail and of the wings edged with orange on a brown ground; the throat white; the head and back gray; the under ſide of the body whitiſh. Such are the principal properties of the bird. It was obſerved by Sloane in Jamaica.

Total length, about ſeven inches and a half; the bill, eight lines and one third; the upper mandible a little protuberant near the middle; the mid toe, eight lines and one third; the alar extent, eleven inches and a quarter; the tail about twenty-three lines.

II.

THE SPOTTED NUTHATCH, *Lath.*

La Sittelle Grivelée, *Buff.*
Sitta Nævia, *Gmel.*
The Wall-creeper of Surinam, *Edw.*

THIS is another American nuthatch, with a hooked bill; but differs from the preceding in ſize, plumage, and climate: it inhabits Dutch Guiana.

The upper ſide of the head and of the body of a dull aſh colour; the ſuperior coverts of the wings, of the ſame colour, but terminated with white; the throat white; the breaſt and all the under ſide of the body cinereous, and more dilute than the upper ſide, with white ſtreaks ſcattered on the breaſt and ſides, which forms a ſort of ſpeckling; the bill and legs brown.

Total length, about ſix inches; the bill, an inch; the *tarſus*, ſeven lines and a half; the mid toe, eight or nine lines, longer than the hind toe whoſe nail is the ſtrongeſt; the tail, about eighteen lines, conſiſting of twelve nearly equal quills; exceeds the wings thirteen or fourteen lines.

THE

CREEPERS.

Les Grimpereaux, *Buff.*

WE have already treated of ſeveral creeping birds, the nuthatches and titmice: we ſhall ſee others in the ſequel, ſuch as the woodpeckers; but the appellation of *creepers* is appropriated to the genus which we are now to conſider. They creep very nimbly on trees, both in aſcending and deſcending; both on the upper and the under ſide of the branches: they run ſwiftly along beams, claſping the edge with their little feet. They are diſtinguiſhed from the woodpeckers by their bill and tongue; and from the titmice by the greater length of their bill; and from the nuthatches by its more ſlender and hooked form; and accordingly they do not ſtrike the bark with it, like theſe other birds.

Many foreign ſpecies of creepers reſemble much the humming birds; by their diminutive ſize, by the rich colours of their plumage, by their ſlender incurvated bill, only it is of a more lengthened and ſharper form, while that of the humming

humming bird is of an equal thickneſs throughout, or even ſlightly inflated at the tip: the legs of the creepers are ſhorter, their wings longer, and their tail contains twelve quills, though that of the humming birds has only ten: and, laſtly, the tongue of the creepers is not, like that of the humming birds, compoſed of two cylindrical half tubes, which, joined together, form an entire tube, and is properly an organ of reſpiration, and more analogous to the feeler of an inſect than the tongue of a bird.

The genus of creepers is alſo ſpread through a wider extent than that of the humming birds. Theſe ſeem peculiar to the continent of America, and ſeldom venture farther than the ſouthern parts of Canada; and at that latitude the breadth of the ocean is too vaſt to be traverſed by a little inſect-bird. But the creeper of Europe penetrating to Denmark, or even beyond, thoſe of Aſia and America probably advanced alſo to the north, ſo that an eaſy communication might be found from one continent to the other.

As the creepers live upon the ſame inſects with the woodpeckers, the nuthatches, and the titmice, and cannot, from the defect of their bill, extract the inſects lodged under the bark, they follow thoſe birds, which they make their providers,

providers, and dexterouſly ſnatch the little prey. And ſince inſects are their ſole ſubſiſtence, we may readily ſuppoſe that the ſpecies are more prolific and varied in hot climates, where ſuch proviſion abounds, than in cold or temperate climates, which are leſs favourable to the multiplication of inſects. This is an obſervation of Sonnerat, and it correſponds to facts.

It is a general remark, that the plumage of young birds is not ſo bright as that of adults; but the difference is more ſtriking in the brilliant tribes of the creepers, the humming birds, and other ſmall birds that inhabit the immenſe foreſts of America. Bajon informs us, that the colours of theſe are formed very gradually, and do not aſſume their luſtre till after a number of moultings. He adds, that the females are ſmaller than the males, and inferior in beauty*.

Whatever analogy ſubſiſts between the creepers of the old and of the new continent, they are yet diſtinct; and I have no doubt but, in time, more important differences will be found both in their exterior appearance and in their natural habits†.

* *Memoires pour ſervir à l'Hiſtoire de Cayenne*, p. 257.

† In Senegal, according to Adanſon, there are many ſpecies of birds, of which the females are as brilliant as the males.

THE
COMMON CREEPER.

Le Grimpereau, *Buff.*
Certhia Familiaris, *Linn. Gmel. Mull. & Brun.*
Certhius Minor, *Frif.*
Falcinellus Arboreus Noftras, *Klein.*
Ifpida Caudâ Rigidâ, *Kram.*
Certhia, *Will. Ray & Briff.**

LITTLE animals are commonly the moft agile. The creeper is nearly as fmall as the crowned wren, and accordingly is perpetually in motion: but the fcene of its activity is extremely limited; it never migrates, and its ordinary abode is the hole of a tree. From this it emerges in purfuit of the infects harboured in the bark and the mofs; and there the female breeds and hatches †. Belon afferts, and almoft all the naturalifts have repeated it, that the creeper has about twenty eggs; but he certainly confounded it with the titmice. For my own part,

* In Greek, Κερθιος, Κερθια, Κερδιος; Arift. *Hift. Anim.* lib. ix. 17. In Italian, *Cerzia Cenerina*, *Picchio Pafferino*, *Rampichino*: in German, *Baum-lauffer* (tree-runner), *Rindenkleber* (bark-climber), *Hirngrille* (brain-cricket): in Danifh, *Træ Pikke*, *Licheften*: in Swedifh, *Krypare*

† Frifch fays, that it defends itfelf ftoutly againft the nuthatch, when invaded.

part, I am confident, both from my own obſervations and thoſe of many naturaliſts*, that the hen generally lays only five eggs, and ſeldom or never above ſeven: they are cinereous, with points and ſtreaks of a deeper colour, and the ſhell is pretty hard. It is obſerved that the hatch is begun early in the ſpring, which is very probable, ſince the bird is neither obliged to conſtruct its neſt nor to migrate.

Friſch aſſerts, that they ſearch for inſects on walls; but ſince he was not acquainted with the real wall-creeper, and did not recognize it in Geſner's deſcription, though diſtinctly characteriſed, he probably confounded here the two ſpecies, eſpecially as the common creeper is recluſe, and lives chiefly in the woods. One was brought to me in the month of January 1773, which had been ſhot on an acacia in the king's garden; but it was regarded as a curioſity, and the people who worked there the whole year told me that they very ſeldom ſaw theſe birds. Nor are they common in Burgundy or Italy †, though frequent in England ‡: they are found alſo in Germany, and as far as

* Salerne, Lottinger, Ginnani.

† Gerini.

‡ Willughby.

Den-

Denmark, as I have already remarked. They have a weak cry, which is very ſhrill and very common. They generally weigh five drachms Engliſh, and appear larger than they really are, becauſe their feathers are not laid regularly upon one another, but briſtled and diſordered, and they are alſo very long.

The throat of the creeper is pure white, but generally aſſumes a ruſty tint, which is always deeper on the flanks and the remote parts (ſometimes all the under ſide of the body is white), the upper ſide variegated with rufous, with white, and with blackiſh; and theſe colours vary in their brightneſs and intenſity: the head is of a darker caſt; the ring about the eyes, and the eyebrows, white; the rump rufous; the quills of the wings brown, the three firſt edged with gray, the fourteen following marked with a whitiſh ſpot, which forms on the wing a tranſverſe bar of the ſame colour; the three laſt marked near the tip with a black ſpot between two white ones: the bill is brown above, and whitiſh below; the legs gray; the ground of the feathers deep cinereous.

Total length, five inches; the bill, eight lines, ſlender, hooked, contracting gradually, and terminating in a point: the throat is wide, ſays Belon; the noſtrils very oblong, half covered

vered by a convex membrane, without any ſmall feathers; the tongue pointed and cartilaginous at the tip, ſhorter than the bill; the *tarſus*, ſeven lines; the mid toe ſeven lines and a half, the lateral toes adhering to the middle one by their firſt phalanx; the hind nail the ſtrongeſt, and even longer than its toe; the nails in general very long, hooked, and calculated for climbing; the alar extent, about ſeven inches; the tail twenty-four lines, according to Briſſon, and twenty-eight, according to Willughby: I have found it to be twenty-ſix: it conſiſts of twelve tapered quills*, the longer ones laid over the ſhorter, which makes the tail appear narrow; they are all pointed at the tip, and the extremity of the ſhaft is worn as in the woodpeckers; but being leſs ſtiff than in theſe birds, it exceeds the wings twelve lines: the wings conſiſt of ſeventeen quills; what is generally reckoned the firſt, and which is very ſhort, ought not to be reckoned among the quills.

The *æſophagus*, two inches; the inteſtines, ſix; the gizzard muſcular, lined with a membrane which is not eaſily detached, and contained portions of inſects, but not a ſingle

* Briſſon, Willughby, and Linnæus reckon only ten quills; but their ſubjects muſt have been incomplete, for I have counted twelve, as well as Pennant and Mœhring.

pebble

pebble or fragment of a ſtone: there were ſlight traces of a *cæcum*, but no gall bladder. [A]

[A] Specific character of the Common Creeper, *Certhia Familiaris*: "It is gray, below white; its wing quills brown, ten of them marked with a white ſpot." It is found likewiſe in America.

VARIETY OF THE CREEPER.

THE GREAT CREEPER. It differs only in ſize; its œconomy, its plumage, and its ſtructure are the ſame as in the common creeper: it ſeems however leſs ſhy and cautious; for Belon mentions the ordinary kind as difficult to catch; but Klein relates, that he once caught one of the great creepers running on a tree.

THE

No. 135

FIG 1 THE SMALL CREEPER FROM THE ISLE OF FRANCE.

FIG. 2. THE SMALL CREEPER FROM GUYANA.

N.o 136

THE WALL CREEPER

THE

WALL CREEPER.

Le Grimpereau de Muraille, *Buff.*
Certhia Muraria, *Linn. Gmel. & Scop.*
Certhia Muralis, *Briſſ.*
Picus Murarius, *Ray, Will. & Kram.*
The Wall-creeper, or Spider-catcher, *Will. & Edw.**

ALL the motions that the preceding performs on trees, this performs on walls; it lodges there, and there it climbs, hunts, and breeds: by walls, I mean not only thoſe built by man, but thoſe formed by nature, the huge perpendicular rocks †. Kramer remarks, that theſe birds prefer the haunts of the tombs, and depoſit their eggs in human ſculls. They fly flapping their wings like the lapwing; and, though they are larger than the common creepers, they are equally lively and active.

* In German, *Mauer-ſpecht* (wall-ſpight), *Kletten-ſpecht* (creeper-ſpight): in Daniſh, *Scopoli:* in Poliſh, *Dzieciot Murowy.*

† In Turin it is called the *mountain woodpecker*; and Schwenckfeld ſays, that it is commonly ſeen in citadels built on mountains.

Flies, ants, and particularly ſpiders, are their uſual food.

Belon ſuppoſed this ſpecies peculiar to Auvergne; but it occurs in Auſtria, Sileſia, Switzerland, Poland, Lorraine, and particularly the part bordering on Germany, and even in England*, according to ſome, though others regard it as at leaſt very rare. On the contrary, it is common in Italy, near Bologna and Florence; but much leſs frequent in Piedmont.

It is chiefly in winter that theſe birds appear near dwellings; and if we believe Belon, they are heard flying at a great diſtance in the air, deſcending from the mountains to lodge on the walls of cities. They keep ſingle, or at leaſt by two and two, like moſt birds that feed on inſects, and though ſolitary, they are neither weary nor melancholy: ſo certain it is, that cheerfulneſs depends more on the original diſpoſition than on the enlivening influence of ſociety!

In the male, there is a black mark under the throat, which extends to the fore part of the neck, and diſtinguiſhes the ſex: the upper ſide of the head and body is of a pleaſant aſh co-

* Edwards thinks, with Ray and Willughby, that it never viſits England, at leaſt he never ſaw it there.

lour,

lour, the under ſide of a deeper caſt; the ſmall ſuperior coverts of the wings, roſe colour; the great ones blackiſh edged with roſe colour; the quills terminated with white, and bordered from their baſe to the middle with roſe colour, which grows more dilute, and almoſt vaniſhes on the quills neareſt the body; the five firſt marked on the inſide with two ſpots of white more or leſs pure, and the nine following with a ſingle fulvous ſpot; the ſmall inferior coverts next the margin roſe coloured, the others blackiſh: the quills of the tail blackiſh, the four mid ones tipt with dirty gray, and the two outer pairs with white; the bill and legs black.

In the female, the throat is whitiſh. A ſubject which I obſerved had, under its throat, a broad mark of light gray, which deſcended on the neck, and ſent off a branch to each ſide of the head. The female deſcribed by Edwards was larger than the male deſcribed by Briſſon. In general, this bird is of a ſize between that of the blackbird and of the ſparrow.

Total length, ſix inches and two-thirds; the bill, fourteen lines; and ſometimes twenty according to Briſſon; the tongue very pointed, broader at the baſe, terminated by two appendices; the *tarſus*, ten or twelve lines; the toes

 diſdifpofed three before and one behind; the mid one nine or ten lines, the hind one eleven; and the chord of the arc formed by the nail alone is fix lines; in general all the nails are long, narrow, and hooked; the alar extent, ten lines; the wings confift of twenty quills, according to Edwards, and of nineteen, according to Briffon; and both include the firft, which is very fhort, and ought not to be reckoned a quill; the tail twenty-one lines, confifting of twelve quills nearly equal; it exceeds the wings fix or feven lines.

Belon pofitively afferts, that this bird has two toes before and two behind; but he alfo fays, that the tail of the common creeper is fhort. The fource of both errors is the fame: that naturalift confidered thefe birds as related to the woodpeckers, and he afcribed thofe characters without examining narrowly. Analogy, which fo often conducts to great difcoveries, frequently milleads in the detail of obfervation. [A]

[A] Specific character of the Wall-creeper, *Certhia Muraria*: "It is cinereous, with a fulvous fpot on the wings."

FOREIGN

FOREIGN BIRDS

OF THE ANCIENT CONTINENT, WHICH ARE RELATED TO THE CREEPERS.

I SHALL term theſe birds *Soui-mangas*, the name which they receive in Madagaſcar. After theſe, I ſhall range the birds of the new continent, which bear ſome analogy to the creepers, but whoſe habits and œconomy are very different; and I ſhall prefer the Indian appellation *guit-guit*, as more expreſſive than any abſtract artificial term. In general the creepers, and ſoui-mangas, have their bill proportionally longer than the guit-guits, and their plumage at leaſt as beautiful, and even equal to that of the moſt brilliant of the humming birds. The colours are the ſofteſt, the richeſt, the moſt dazzling; all the tints of green, of blue, of orange, of red, of purple, heightened by the contraſt of various ſhades of brown and gloſſy black. We cannot enough admire the glow of theſe colours, their ſparkling luſtre, their endleſs variety, even in the dried ſpecimens which decorate our cabinets. Nature would ſeem to have formed the feathers of precious ſtones; of the ruby, of the emerald, the amethyſt, and the topaz. How enchanting, could we view the birds themſelves!

their plumage in all its freſhneſs, animated by the breath of life, embelliſhed by all that dazzles in the magic of the priſm, changing its reflections with each quick movement, and darting new colours or new flames. To ſtudy nature in her minute, as in her grand productions, we ought to contemplate her in the ſtate of freedom, before the hand of man has interfered.

There are many ſoui-mangas living with the Dutch bird-catchers at the Cape of Good Hope: the only food offered is ſugared water: the flies, which abound in that climate and torment Dutch cleanlineſs, ſupply the reſt. Theſe birds are alert in ſeizing them, and none eſcape that enter their volery. This additional food ſeems neceſſary to their ſupport: for they ſoon die on board ſhips, where there are fewer inſects. The Viſcount Querhoënt, to whom we are indebted for theſe remarks, could never keep them alive above three weeks.

I.

THE SOUI-MANGA.

Certhia Soui-Manga, *Gmel.*
Certhia Madagafcarenfis Violacea, *Briff.*
The Violet Creeper, *Lath.*

THE head, the throat, and all the anterior part, are of a fine brilliant green, with a double collar of violet and chefnut: but thefe colours are not uniform or permanent: the light which plays among the webs of the feathers changes inceffantly its fhades, from gold green to deep blue: on each fide, below the fhoulder, there is a fpot of fine yellow: the breaft is brown; the reft of the under fide of the body, faint yellow; the reft of the upper fide of the body, dufky olive; the great coverts and quills of the wings brown, edged with olive; thofe of the tail black, edged with green, except the outermoft, which is partly brown gray: the following one is terminated with the fame colour: the bill and legs are black.

The female is rather fmaller, and much inferior in beauty: it is olive brown above, olive bordering on yellow below; in other refpects fimilar to the male, but inferior in luftre. It is nearly the fize of the common wren.

Total length, about four inches; the bill, nine lines; the *tarſus*, above ſix lines; the middle toe five lines and an half, larger than the hind one; the alar extent, ſix inches; the tail fifteen lines, conſiſting of twelve equal quills, and exceeds the wings ſeven or eight lines.

WE may refer, as a variety cloſely related to this ſpecies, the ſoui-manga from the iſland of Lucon, which I ſaw in the excellent cabinet of M. Mauduit. Its neck and throat are ſteel coloured, with reflections of green, blue, violet, &c. and ſeveral collars, which the brilliant play of theſe reflections ſeems to multiply. We may however diſtinguiſh four that are more conſtant; the lower blackiſh violet, the next cheſnut, then brown, and laſtly yellow: there are two ſpots of yellow below the ſhoulders: the reſt of the under ſide of the body is olive gray; the upper ſide deep green, with reflections of blue, violet, &c. and the quills of the wings, the quills and coverts of the tail, are brown of various intenſity, with a greeniſh gloſs.

Total length, a little leſs than four inches; the bill, ten lines; the *tarſus*, ſeven; the hind nail

nail the ſtrongeſt; the tail fifteen lines, and ſquare, and exceeds the wings ſeven lines.

II.

THE RED-BREASTED PURPLE-CHESNUT SOUI-MANGA.

Le Soui-Manga Marron Pourpré à Poitrine Rouge, *Buff.*
Certhia Sperata, *Linn.* & *Gmel.*
Certhia Philippenſis Purpurea, *Briſſ.*
The Red-breaſted Creeper, *Buff.*

SEBA ſays that the ſong of this bird reſembles that of the nightingale: the head, the throat, and the fore part of the neck, are variegated with fulvous and gloſſy black, changing into violet blue: the upper ſide of the neck and of the body, in the fore part, purpliſh cheſnut; and, in the hind part, violet, changing into gold green: the ſmall coverts of the wings the ſame; the middle ones brown, terminated with purpliſh cheſnut; the reſt of the under ſide of the body, olive yellow; the quills and great coverts of the wings brown, edged with rufous; the quills of the tail blackiſh, with ſteel reflections, and edged with violet, changing into gold green; the bill black above (yellow, according to to Seba), whitiſh below ; the legs brown (yellowiſh, according to Seba), and the nails long.

The female differs from the male, being olive green above, and olive yellow below ; with the quills of the tail blackiſh, and the four lateral pairs tipt with gray. Theſe birds are rather ſmaller than the common creepers.

Total length, four inches ; the bill, eight lines ; the *tarſus*, ſix; the mid toe, five ; the hind one rather ſhorter ; the alar extent, ſix inches ; the tail an inch, conſiſting of twelve quills, and exceeding the wings three lines. [A]

[A] Specific character of the *Certhia Sperata* : "It is purple ; below crimſon ; its head, throat, and rump, purple.'

Varieties of the preceding.

1. The Little Creeper, or Soui-Manga*, the little brown and white creeper, or honey thief, of Edwards, reſembles ſo much the preceding, that I muſt conſider it as a variety

* Certhia Puſilla, *Linn. & Gmel.*
Certhia Indica, *Briſſ.*

Specific character: "It is brown ; below white ; its eyebrows bright white ; its tail quills brown ; the outermoſt tipt with white."

riety of age, its plumage not being formed, and only beginning to aſſume the reflections. It is white below, and brown above, with ſome reflections of copper colour: it has a brown ſtreak between the bill and the eye, and white eyebrows: the quills of the wings are of a deeper brown than the back, and edged with lighter colours; the quills of the tail blackiſh, the outermoſt terminated with white; the bill and legs, brown. Edwards ſays that it is only half as large as the European creeper.

Total length, three inches and an half; the bill, eight or nine lines; the *tarſus*, five or ſix; the mid toe five, rather longer than the hind one; the tail thirteen lines, conſiſting of twelve equal quills, and exceeds the wings three or four lines.

2. The Creeper, or Soui-Manga*, with a violet throat and red breaſt, brought from New Guinea by Sonnerat. Its back, and the ſmall quills of the wings, ſnuff-colour; its

* Certhia Senegalenſis, *Linn. & Gmel.*
Certhia Senegalenſis Violacea, *Briſſ.*
The Senegal Creeper, *Lath.*

Specific character: "It is violet black; its top and its throat gold green: its breaſt crimſon."

rump

2

rump and tail like burniſhed ſteel, and verging on greeniſh; and the inferior coverts of the tail are of a dirty green. It is alſo a native of the Philippine Iſlands.

III.

THE RED-BREASTED VIOLET SOUI-MANGA.

Certhia Philippina, *Linn. Gmel. & Briſſ.*
The Philippine Creeper, *Lath.*

ITS leading colour is violet; and on this ground the brighter tints of the anterior parts appear with advantage: on the throat and the upper ſide of the head is a brilliant gold green, heightened with copper reflections: on the breaſt and the fore part of the neck is a beautiful ſhining red the only colour which is ſeen in thoſe parts, when the feathers are quite regular and compoſed: each of the feathers, however, conſiſts of three different colours, black at its origin, gold green at its middle, and red at its extremity; a proof (a thouſand ſuch may be adduced) that to deſcribe the tints of the feathers is not ſufficient to give an accurate idea

idea of the colours of the plumage. All the quills of the tail and of the wings, the great superior coverts of the wings, and their inferior coverts, brown: the thighs are of a mixed caſt, in which the brown ſeems melted with the violet: the bill is black, and the legs blackiſh. This bird is nearly of the ſize of the golden-crowned wren. It is found in Senegal.

Total length, five inches; the bill, ten lines; the *tarſus*, ſeven lines; the mid toe five lines and an half, rather longer than the hind one; the alar extent, ſeven inches and one third; the tail twenty-two lines, and compoſed of twelve equal quills: it exceeds the wings ten lines. [A]

[A] Specific character of the *Certhia Philippina*: "The two middle tail quills are very long; the body grayiſh, with a greeniſh caſt, below yellowiſh white."

IV.

THE PURPLE SOUI-MANGA*.

IF this bird had been of a varying gold green on the head and under the throat, and red, in-

* The Purple Indian Creeper, *Edwards*. This author ſays that this bird has the tongue of the colibri; that is, divided at the tip into many filaments. Edwards ſeems therefore not to have been well acquainted with the true ſtructure of the tongue of the colibri.

ſtead of green and yellow, on the breaſt, it would have been almoſt exactly like the preceding; or, at leaſt, it would have been more analogous than the collared ſoui-manga, which has not a ſhade of purple in its plumage. I cannot conceive why Briſſon conſiders the latter, and the purple creeper of Edwards, as preciſely the ſame, only with different names.

V.

THE COLLARED SOUI-MANGA.

Le Soui-Manga à Collier, *Buff.*
Certhia Chalybea, *Linn* & *Gmel.*
Certhia Torquata Capitis Bonæ Spei, *Briſſ.*
The Collared Creeper, *Lath.**

THIS ſpecies, which comes from the Cape of Good Hope, bears ſome analogy to that of the violet ſoui-manga: its head is likewiſe of a gold green, waving with roſe copper; and this gold green extends over the throat, the head, and all the upper ſide of the body; it borders alſo the intermediate quills of the tail,

* Briſſon, Linnæus, Gmelin, and Latham agree to refer the *Purple Indian Creeper* of Edwards to this bird.

which

which are of a gloſſy black; only it is not changeable on the ſuperior coverts. The breaſt is marked with red, as in the violet ſoui-manga, only confined to a narrower ſpace, and not raiſed ſo high, and forming a ſort of cincture whoſe upper edge is contiguous to the collar of blue ſteel colour, waving with green, and about a line in breadth: the reſt of the under ſide of the body is gray, with ſome yellow ſpeckles on the top of the belly, and on the flanks: the quills of the wings are of a brown gray: the bill is blackiſh, and the legs entirely black. The bird is nearly of the ſize of the violet ſoui-manga, but differently proportioned.

Total length, four inches and a half; the bill, ten lines; the *tarſus*, eight lines and a half; the mid toe ſix lines, and nearly equal to the hind toe; the alar extent, ſix inches and a half; the tail eighteen lines, conſiſting of twelve equal quills, and exceeding the wings nine lines.

The female, according to Briſſon, differs from the male, the under ſide of its body being of the ſame colour with the upper ſide, only there are yellow ſpeckles on the flanks: according to others, it has alſo a red cincture, but which falls lower than in the male, and all its other colours

are

are fainter: admitting this, we may regard as the female the ſoui-manga obſerved at the Cape of Good Hope, by the Viſcount Querhoënt, in January 1774. In that bird the throat was brown gray, variegated with green and blue; the breaſt decorated with a flame-coloured cincture; the reſt of the under ſide of the body, white gray; the head and all the upper ſide of the body brown gray, variegated with green on the back, and with blue at the origin of the tail; the wings light brown, with a coat of gold yellow; the quills of the tail, blackiſh; the bill and legs, black. The Viſcount ſubjoins, that this bird ſings prettily; that it lives on inſects and the juices of flowers; but that its throat is ſo narrow that it cannot ſwallow the larger common flies. Is it not then probable that this was a young bird, and that the true female of the ſoui-manga is the ſame with Briſſon's creeper from the Cape of Good Hope*, which is uniformly of a brown gray, deeper above and lighter below, the colour which borders the tail and the wings? Their dimenſions alſo correſpond, and both are brought from the African promontory; but time and obſervation will aſcertain the point.

* Certhia Capenſis, *Linn.* edit. xiii.

Laſtly,

Laſtly, we may reckon, as a female of this ſpecies, or as one of its varieties, the Philippine creeper of Briſſon*, whoſe plumage, uniform and without brilliancy, indicates a female, and the middle quills of whoſe tail are edged with a ſhining black, waving with gold green, like the quills of the tail of the collared ſoui-manga; but, in this female, the reflections are much leſs bright: it is of a greeniſh brown above, with a ſulphur caſt below; the quills of the wings brown, edged with a lighter colour, and the lateral ones of the tail blackiſh, terminated with dirty white.

If the creepers of the Eaſt Indies, like thoſe of America, require ſeveral years to form their plumage, and if the rich colours be not aſſumed till after a number of moultings, we need not be ſurpriſed that ſo many varieties are found.

Total length, four inches nine lines; the bill, an inch; the *tarſus*, ſix lines and a half; the mid toe, five lines and a half; the hind one

* Certhia Philippina, *Linn.* edit. xiii. (See art. iii. the note.) I know not on what foundation Linnæus gives to this ſpecies two long quills in the middle of the tail: if he ſaw an individual ſo formed, it muſt have been a young one, or an old one in moult, or a female. But I am diſpoſed to think that Linnæus never ſaw this bird, ſince he does not deſcribe it, and adds nothing to what others have ſaid.

almoſt as long; the alar extent, ſix inches and a quarter; the tail fifteen lines, conſiſting of twelve equal quills, and projecting five lines beyond the wings. [A]

[A] Specific character of the *Certhia Chalybea*: "It is of a gloſſy green, its breaſt red, a ſteel-coloured bar before."

VI.

THE PURPLE-BREASTED OLIVE SOUI-MANGA.

Le Soui-manga Olive à Gorge Pourpre, *Buff.*
Certhia Zeylonica, *Linn. & Gmel.*
Certhia Philippenſis Olivacea, *Briſſ.*
The Ceyloneſe Creeper, *Lath.*

THE moſt conſpicuous colour of its plumage is a deep and very brilliant violet, which ſpreads below the neck, and on the throat: the reſt of the under ſide of its body is yellow; all the upper ſide, including the ſuperior coverts of the wings, of a dull olive, and the ſame colour borders the quills of the tail and of the wings, and alſo their great coverts, of which brown

brown is the prevailing colour; the bill is black, and the legs deep cinereous.

Poivre brought this bird from the Philippines; it is nearly of the ſize of the common wren.

Total length, four inches; the bill, nine or ten lines; the *tarſus*, ſix lines; the mid toe five lines; the hind toe rather ſhorter; the alar extent, ſix inches; the tail fourteen lines, conſiſting of twelve equal quills, and exceeds the wings ſix lines. [A]

If the bill were not ſhorter and the tail longer, I ſhould regard the Madagaſcar creeper of Briſſon * as the female of the ſoui-manga of this article; it is at leaſt an imperfect or degenerated variety: all the upper part of the body, including the coverts of the wings, is of a dull olive green, but darkeſt on the crown of the head, and the ſame colour borders alſo the quills of the wings and of the tail: all theſe quills are brown; the orbits are whitiſh; the

[A] Specific character of the *Certhia Zeylonica*: "It has a green cap; its back is ferruginous, its belly yellow, its throat and rump azure."

* Certhia Olivacea, *Linn. & Gmel.*
The Olive Creeper. *Lath.*
Specific character: "It is olive, below brown, its orbits whitiſh."

throat and the under ſide of the body, dun gray; the legs entirely brown: the bill is blackiſh. It is nearly as large as the common creeper.

Total length, four inches; the bill, ſix or ſeven lines; the *tarſus*, ſeven lines; the middle toe five lines and a half, and the hind one rather ſhorter; the alar extent, ſix inches and a half; the tail nineteen lines, conſiſting of twelve equal quills, and exceeds the wings eight lines.

There is a bird in the Philippine Iſlands which may alſo be conſidered as a variety of the ſame ſpecies*: the under ſide of the body is of a pretty ſhade of dun gray, and the upper ſide yellowiſh; the breaſt darker: there is a deep violet bar, which riſes from the throat and deſcends along the neck: the coverts of the wings are of a ſteel colour, and the ſame borders the quills of the tail, the reſt of which are blackiſh; the lateral ones are terminated with dirty white; the quills of the wings, brown; the bill ſtronger than in the other creepers, and the tongue terminated by two threads, according to Linnæus;

* Certhia Currucaria, *Linn.* & *Gmel.*
Certhia Philippenſis Griſea, *Briſſ.*
The Gray Creeper, *Lath.*

Specific character: "It is olive, below yellowiſh, its tail quills equal."

the

the bill and legs black: it is ſmaller than the common creeper.

Total length, four inches and two-thirds; the bill, nine lines; the *tarſus*, ſix lines and a half; the mid toe, five lines and a half; the hind toe rather ſhorter; the alar extent, ſix inches and a quarter; the tail fifteen lines, conſiſting of twelve equal quills, and projecting five lines beyond the wings.

* Laſtly, we may reckon the little creeper from the Philippines as a ſecondary variety of the preceding. It is always brown gray above, and yellow below; it has a violet collar; the quills of the wings are brown gray, like the upper ſide of the body; thoſe of the tail are deeper brown; the two outermoſt pairs are terminated with dirty white; the bill and the legs are blackiſh. This bird is much ſmaller than the former, which it reſembles much in regard to plumage, and perhaps it is the leaſt of all the ſoui-mangas known in the ancient continent; which affords a preſumption that it is a young one.

Total length, three inches and two-thirds; the bill, nine lines; the *tarſus*, ſix lines; the

* Certhia Jugularis, *Linn. & Gmel.*
Certhia Philippenſis Minor, *Briſſ.*
Specific character: "It is grayiſh, below yellow, its throat violet, the two outermoſt quills of the tail tipt with yellow."

 mid

mid toe four lines and a half; the hind one rather ſhorter; the alar extent, five inches and two-thirds; the tail fifteen lines, conſiſting of twelve equal quills, and projects five lines beyond the wings.

VII.

THE ANGALA DIAN, *Buff.*

Certhia-Lotenia, *Linn. & Gmel.*
Certhia Madagaſcarenſis Viridis, *Briſſ.*
Loten's Creeper, *Lath.*

THIS bird has alſo a collar, a line and a half broad, and of a bright ſcarlet; the ſmall ſuperior coverts of the ſame hue; the throat, the head, the neck, all the upper ſide of the body, and the middle coverts of the wings, of a brilliant gold green: there is a ſtreak of gloſs black between the noſtril and the eye; the breaſt, the belly, and all the under ſide of the body, of the ſame black, and alſo the quills of the tail and of the wings, and the greater coverts of the wings: but theſe large coverts, and the quills of the tail, are bordered with gold green: the bill is black, and ſo are the legs. [A]

[A] Specific character: "It is blue, with a gold red ſtripe on its breaſt; its ſtraps black."

Adanſon

Adanſon ſuſpects that the bird which Briſſon conſiders as the hen angala is only a young one of the ſame ſpecies before its firſt moulting. " This ſeems to appear," he ſubjoins, " from the number of birds of this kind, and very ſimilar to it, which are found at Senegal, and of which the females are exactly like the males: but the young ones have a great intermixture of gray, which they loſe not till they drop their feathers *."

The angala is almoſt as large as the epicurean warbler: it gives its neſt the ſhape of a cup, like the canary and chaffinch, and ſcarcely uſes any other materials than the down of plants. It lays generally five or ſix eggs: but it is often driven from its hatch by a ſort of large voracious ſpider, which ſeizes the brood, and ſucks the blood †.

The bird which Briſſon regards as a female, and Adanſon as a young one, is of a dirty white,

* I doubt not that M. Adanſon ſaw at Senegal numbers of females like their males, ſince he avers it; but we muſt not thence draw any general rule for all the birds of Africa and of Aſia: the gold pheaſant of China, the peacock, many ſpecies of turtles, of ſhrikes, of parrakeets, &c. found in Africa, are proofs to the contrary.

† Supplement de l'Encyclopedie, au mot *Angala*.

interſperſed with black ſpots on the breaſt, and the reſt of the under ſide of the body, inſtead of an uniform gloſſy black; and its wings and tail are alſo of a leſs brilliant black.

Total length, five inches and a quarter; the bill, fourteen lines; the *tarſus*, eight lines; the middle toe ſix lines and a half, and larger than the hind one; the alar extent, eight lines; the tail nineteen lines, and conſiſting of twelve equal quills: it exceeds the wings ſix or ſeven lines.

VIII.

THE IRIS SOUI-MANGA.

Le Soui-Manga de Toutes Couleurs, *Buff.*
Certhia Omnicolor, *Linn. & Gmel.*
The Green-gold Creeper, *Lath.*

OUR knowledge of this bird is very ſcanty: it comes from Ceylon: its plumage is green, tinged with all the rich colours, among which that of gold ſeems to predominate. Seba ſays, that its young often become the prey of large ſpiders; a danger to which the angala alſo is liable, and even all the ſmall birds that breed in climates

climates inhabited by thofe formidable infects, and have not fkill fufficient to guard the neft from their intrufions.

If we judge from the figure which Seba has given, this bird is feven or eight inches in total length; its bill, about eighteen lines; the tail, two inches and a quarter: in fhort, it appears to be the largeft of the foui-mangas. [A]

[A] Specific character of the *Certhia Omnicolor*: "It is green, mixed with all forts of colours."

IX.

THE RED-BREASTED GREEN SOUI-MANGA.

Le Soui-Manga Vert à Gorge Rouge, *Buff.*
Certhia Afra, *Linn.* & *Gmel.*
The Red-breafted Green Creeper, *Edw.*
The African Creeper, *Lath.*

SONNERAT, who brought this bird from the Cape of Good Hope, tells us that it fings as well as the nightingale, and that its voice is even fofter. Its throat is of a fine carmine; its belly, white; the head, the neck, and the anterior part of the wings, of a fine gold green, and

and ſilvery; the rump, ſky blue; the wings and tail, ſnuff brown; the bill and legs, black.

Total length, nearly four inches and two-thirds; the bill, an inch; the tail eighteen lines, and exceeds the wings about thirteen lines. [A]

[A] Specific character of the *Certhia Afra*: "It is green, its belly white, its breaſt and rump ſky blue."

X.

THE BLACK, WHITE, AND RED INDIAN CREEPER, OR SOUI-MANGA.

Certhia Cruentata, *Linn. & Gmel.*
Certhia Bengalenſis, *Briſſ.*
The Red-ſpotted Creeper, *Lath.*

THIS is the appellation which Edwards beſtows on this bird, which is nearly of the ſize of the crowned wren. The white is ſpread on the throat, and all the lower part, without exception; the black, on the upper part: but on this dark ground, which is ſlightly gloſſed with blue, there are ſcattered four beautiful ſpots of bright red; the firſt on the crown of the head, the ſecond behind the neck, the third on the back, and the fourth on the ſuperior coverts of the

the tail. The quills of the tail and of the wings, the bill and the legs, are black.

Total length, three inches and a quarter; the bill, five or ſix lines; the *tarſus*, five lines; the mid toe, four or five lines; the hind toe rather ſhorter; the tail about an inch, conſiſting of twelve equal quills, and exceeding the wings by five or ſix lines. [A]

[A] Specific character of the *Certhia Cruentata*: "It is dark blue; below white; its top, its neck, its back, and its rump, red."

XI.

THE BOURBON SOUI-MANGA.

Le Soui-Manga de l'Iſle Bourbon, *Buff.*

I ASSIGN no particular name to this bird, becauſe I ſuſpect that it is either a female, or a young male whoſe plumage is not fully ripened. It appears the neareſt related to the proper ſoui-manga, or violet creeper. The upper ſide of the head and body, greeniſh brown; the rump, olive yellow; the throat, and all the upper ſide of the body, of a confuſed gray, which aſſumes a yellowiſh caſt near the tail; the flanks, rufous; the quills of the tail, blackiſh;

ifh; thofe of the wings blackifh, edged with a lighter colour; the bill and legs black.

The dimenfions are nearly the fame with thofe of the violet creeper.

THE

LONG-TAILED SOUI-MANGAS.

WE are acquainted with only three birds in the ancient continent to which this epithet belongs. Seba mentions alfo a female of that kind, which had not the long tail; whence it would appear that, in fome fpecies at leaft, this is the attribute of the male. And might not feveral males already defcribed have attained the fame character at the proper age or feafon? Many fubjects which have been figured and engraved, are perhaps only females, or young males, or even old ones in moult, when this decoration is loft. For there is no difference between the conformation of the long-tailed foui-mangas and the fhort-tailed ones; and their plumage glows with the fame colours.

I. THE

I.

THE LONG-TAILED VIOLET-HOODED SOUI-MANGA.

Le Soui-Manga à Longue Queue & à Capuchin Violet, *Buff.*
Certhia Violacea, *Linn. & Gmel.*
Certhia Longicauda Minor Capitis Bonæ Spei, *Briff.*
The Violet-headed Creeper, *Lath.*

I SEE no reafon why this bird fhould have been called the little creeper, unlefs that the two middle quills of the tail are not fo long as in the two others; but, if we overlook the tail, this will be found not the fmalleft of the three. It refembles fo ftrongly the purple chefnut fouimanga, or red-breafted creeper, that had it not been larger, and its tail differently formed, I fhould have reckoned them both of the fame fpecies, the one having loft its tail in moulting. The Vifcount Querhoënt faw it in its native climate, near the Cape of Good Hope: he informs us that it conftructs its neft artfully, and ufes no other materials but a filky bur.

The head, the top of the back, and the throat, are of a bright violet, gloffed with green; the fore part of the neck alfo bright violet, but gloffed with blue: the reft of the upper fide of the

the body is of an olive brown, which colour borders the great coverts of the wings, their quills, and thoſe of the tail, which are all brown, more or leſs deep; the reſt of the under ſide of the body orange, which is more vivid on the anterior parts, and ſpreads ſoftening into the diſtant parts. The bird is in a ſlight degree larger than the common creeper.

Total length, above ſix inches; the bill, eleven lines and an half; the legs, ſeven lines and an half; the mid toe ſix lines, and a very little longer than the hind one; the alar extent, ſix inches and one-third; the tail three inches, and conſiſting of ten lateral tapered quills, and two intermediate ones, which project twelve or fourteen lines beyond the lateral ones, and twenty-ſeven lines beyond the wings: theſe two intermediate ones are narrower than the lateral ones, but broader than in the following ſpecies. [A]

[A] Specific character of the *Certhia Violacea*: "The two middle quills of the tail are very long: the body is of a gloſſy violet; the breaſt and belly yellow."

II. THE

II.

THE LONG-TAILED SOUI-MANGA

OF A GLOSSY GOLD GREEN.

Le Soui-Manga Vert Doré Changeant, a Longue Queue, *Buff.*
Certhia Pulchella, *Linn. & Gmel.*
Certhia Longicauda Senegalenſis, *Briſſ.*
Sylvia Verſicolor, *Klein.*
Avis Amboinenſis Diſcolor, *Seba.*
The Beautiful Creeper, *Lath.*

THE breaſt is red; all the reſt of a pretty deep gold green, but glowing and undulating with roſe copper; the quills of the tail blackiſh, edged with the ſame green; thoſe of the tail, and their great coverts, brown; the lower belly mixed with a little white; the bill black, and the legs blackiſh.

This ſpecies comes from Senegal. In the female the upper ſide is greeniſh brown; the under ſide yellowiſh, variegated with brown; the inferior coverts of the tail white, ſprinkled with brown and blue; the reſt as in the male, except a few ſhades.

Total length, ſeven inches and two lines; the bill, eight lines and an half; the *tarſus*, ſeven lines; the mid toe five lines and an half, longer

 than

than the laſt; the alar extent, ſix inches and a quarter; the tail four inches three lines, conſiſting of ten lateral quills, nearly equal, and two intermediate ones, which are very long and narrow, and which project two inches eight lines beyond theſe, and three inches four lines beyond the wings. [A]

[A] Specific character of the *Certhia Pulchella*: "The two middle quills of the tail are very yellow, its body gloſſy green, its breaſt red."

III.

THE GREAT GREEN LONG-TAILED SOUI-MANGA.

Le Grand Soui-Manga Vert à Longue Queue, *Buff.*
Certhia Famoſa, *Linn. & Gmel.*
Certhia Longicauda Capitis Bonæ Spei, *Briſſ.*
The Famous Creeper, *Lath. Syn.*

THIS bird inhabits the Cape of Good Hope, where it was obſerved and kept ſeveral weeks by the Viſcount Querhoënt, who deſcribes it in the following terms: "It is of the ſize of the linnet; its bill, which is a little incurvated, is fourteen

fourteen lines long; it is black, and alſo the feet, which are furniſhed with long nails, particularly the middle and hinder ones: the eyes are black; the upper and under ſides of the body, of a very fine brilliant green (gloſſed with roſe copper, Briſſon adds), with ſome feathers of gold yellow under the wings; the great feathers of the wings and of the tail, of a fine black, gloſſed with violet; the filament of the tail, which is rather more than three inches, is edged with green." Briſſon adds, that on each ſide, between the bill and the eye, there is a ſtreak of velvet black.

In this ſpecies the female has alſo a long tail, or rather a long filament at its tail, but which is ſhorter, however, than in the male, for it projects only two inches and a few lines beyond the lateral quills: the upper ſide of the body and of the head greeniſh brown, mixed with ſome feathers of a fine green; the rump green; the great quills of the wings and of the tail almoſt black, and alſo the filament or two intermediate quills: the under ſide of the body is yellowiſh, with ſome green feathers on the breaſt. [A]

[A] Specific character of the *Certhia Famoſa*: "The two middle quills of its tail are very long; its body of a gloſſy green, yellow under its pinions, the ſtraps black."

IV.

THE CREEPER-BILLED RED BIRD.

L'Oiseau Rouge à Bec de Grimpereau, *Buff.*
Certhia Mexicana, *Gmel.*
Trochilus Coccineus, *Linn.* edit. 6th.
Certhia Mexicana Rubra, *Briss.*
Avicula Mexicana, *Klein & Seba.*
The Red Creeper, *Lath.*

THIS and the three following have been reckoned American birds, and would therefore range with the guit-guits; but as from their conformation, and particularly the length of their bill, they are more related to the soui-mangas, we have placed them between the two tribes. We adopt this plan the more readily, as the climate of these birds rests solely upon the authority of Seba, which naturalists know has little weight, and ever insufficient to balance against the force of analogy. We shall not, however, hurt the received prejudices by changing the name; though that of soui-manga would suit them better.

Red is the prevailing colour in the plumage of this bird, but it has different shades; for the crown of the head is lighter and more brilliant, and that on the rest of the body is deeper. There are some exceptions, however: for the throat and

and the fore part of the neck are green; the quills of the tail and of the wings are terminated with blueiſh; the thighs, the bill, and the legs, of a light yellow.

Its voice is ſaid to be pleaſant. It is a little larger than our creeper.

Total length, about four inches and a half; the bill, ten lines; the *tarſus*, ſix lines; the mid toe five lines, rather longer than the hind one; the tail fourteen lines, conſiſting of twelve equal quills, and projecting about ſeven lines beyond the wings.

I conſider as a variety of this ſpecies the black-headed red bird, which Seba and ſome others after him have referred to New Spain. Its proportions are exactly like thoſe of the preceding: the only apparent difference is in the length of the bill, which is ten lines in the preceding, and only ſeven in this, which would occaſion a difference in the total length. But theſe meaſures are taken from the figure, and therefore liable to error; eſpecially as the original obſerver, Seba, ſeems more ſtruck with its long bill than with that of the other. It is very probable that the deſigner or engraver took the liberty of ſhortening it; and an alteration of three or four lines would bring the two birds to an almoſt perfect identity. There are ſome

 differ-

differences in the plumage, which alone induced me to diſtinguiſh it as a variety.

Its head is of a fine black, and the ſuperior coverts of the wings, gold yellow; all the reſt is light red, except the quills of the tail and wings, which are of a deeper ſhade. The dimenſions preciſely as in the preceding bird. [A]

[A] Specific character of the *Certhia Mexicana*: "It is red; its throat green, the tip of its wing quills blueiſh."

V.

THE CREEPER-BILLED BROWN BIRD.

L'Oiſeau Brun à Bec de Grimpereau, *Buff.*
Certhia Gutturalis, *Linn. & Gmel.*
Certhia Braſilienſis Nigricans, *Briſſ.*
The Green-faced Creeper, *Lath.*

THE bill of this bird is two-ſevenths of the length of the body: the throat and face are of a fine gold green; the fore part of the neck of a bright red; the ſmall coverts of the wings of a brilliant violet; the great coverts and the quills of the wings and of the tail are brown, tinged

tinged with rufous; the middle coverts of the wings, and all the reſt both of the upper and under ſides of the body, blackiſh brown; the bill and the legs black.

This bird is not larger than the epicurean warbler.

Total length, five inches one-third; the bill, one inch; the *tarſus*, ſeven lines and a half; the mid toe ſix inches, and larger than the hind one; the alar extent, eight inches; the tail twenty-one lines, conſiſting of twelve equal quills, and exceeding the wings about ſeven lines. [A]

[A] Specific character of the *Certhia Gutturalis*: "It is blackiſh, its throat gloſſy green, its breaſt purple."

VI.

THE CREEPER-BILLED PURPLE BIRD.

L'Oiſeau Pourpré à Bec de Grimpereau, *Buff.*
Certhia Purpurea, *Gmel.*
Certhia Virginiana Purpurea, *Briſſ & Gerini.*
Avis Virginiana Pœnicea, *Seba & Klein.*
The Purple Creeper, *Lath.*

ALL its plumage, without exception, is of a beautiful uniform purple. Seba applies arbitrarily the name *atototl*, which, in Mexican, ſignifies an aquatic bird; though the preſent is entirely of a different claſs. Seba aſſerts, I know not on what authority, that it ſings pleaſantly: it is rather larger than the epicurean warbler.

Total length, four inches and a half; the bill, above an inch; the *tarſus*, ſix lines and a half; the mid toe five lines and a half, rather longer than the hind toe; the tail fourteen lines, and exceeds the wings ſeven lines. [A]

[A] Specific character of the *Certhia Purpurea*: "It is entirely purple."

THE

AMERICAN GUIT-GUITS.

GUIT-GUIT is an American name, applied to one or two of this tribe, containing the creepers of the New Continent: I ſhall uſe it as a generic appellation. I have already noticed ſome differences that obtain between them and the humming birds: I may add that they neither fly in the ſame manner, nor ſip the nectar of flowers. Yet the creoles at Cayenne frequently confound them; and we ſhould be therefore aware of this circumſtance in reading the relations of travellers.

I am aſſured that the guit-guits of Cayenne never climb upon trees; that they live in flocks with thoſe of their own kind, and alſo with other birds, ſuch as the little tanagres, nuthatches, picucullas, &c. and that they feed not only upon inſects, but upon fruits and even buds.

I.

THE BLACK AND BLUE GUIT-GUIT, *Buff.*

Certhia Cyanea, *Linn. & Gmel.*
Certhia Brafilienfis Cærulea, *Briff.*
Guira-cœreba, *Ray & Will.*
The Black and Blue Creeper, *Edw. & Lath.*

THE face of this beautiful bird is of a brilliant fea green: there is a bar on the eyes of velvet black; the reft of the head, the throat, and all the under part of the body (without exception, according to Edwards), the lower part of the back, and the fuperior coverts of the tail, of an ultramarine blue, which is the only colour that appears when the feathers are regularly difpofed, though each has three colours, according to the remark of Briffon, brown at the bafe, green in the middle, and blue at the extremity; the top of the back, the part of the neck contiguous to the back, and the tail, velvet black: what appears of the wings, when they are clofed, is of the fame black, except a blue bar, which croffes their coverts obliquely: the inner fide of the quills of the wings, and their inferior coverts, are of a fine yellow; fo that the wings, which feem entirely black when at reft, reſt, appear variegated with black and yellow when diſplayed, or in motion. The inferior coverts of the tail are of a dull black (and not blue, as Briſſon repreſents): the bill is black, and the legs ſometimes red, ſometimes orange, ſometimes yellow, and occaſionally whitiſh.

It appears from this deſcription, that the colours of the plumage are ſubject to vary in different ſpecimens: in ſome, the throat is mixed with brown; in others it is black. In general, the diſtribution of the black ſeems the moſt irregular: the blue aſſumes ſometimes a violet tinge.

Marcgrave obſerved, that the eyes are black; that the tongue is terminated by many filaments; that the feathers on the back are ſilky; and that the bird is nearly as large as the chaffinch. He ſaw it in Brazil; but it occurs alſo in Guiana and Cayenne. In the female the wings have a coat of yellowiſh gray.

Total length, four inches and a quarter; the bill, eight or nine lines; the *tarſus*, ſix or ſeven; the mid toe ſix, and a very little longer than the hind toe; the alar extent, ſix inches and three quarters; the tail fifteen lines, conſiſting of twelve equal quills, and exceeding the wings three or four lines. [A]

[A] Specific character of the *Certhia Cyanea*: "It is ſky blue; a bar on the eyes; the ſhoulders, the wings, and the tail, black; its legs brown."

VARIETY of the BLACK AND BLUE GUIT-GUIT *.

THIS variety is found in Cayenne: it differs from the preceding only in the ſhades of the plumage: the head is of a fine blue: there is a bar on the eyes of a velvet black: the throat, the wings, and the tail, are of the ſame black: all the reſt is of a ſhining blue, verging upon violet; the bill black, and the legs yellow: the blue feathers which cover the body, are of three colours, the ſame as in the preceding.

With regard to ſize, it is rather ſmaller, and the tail eſpecially appears ſhorter; which would imply that it is either a young bird, or an adult that has not recovered from moulting; but the alar extent is greater, which precludes this ſuppoſition.

It conſtructs its neſt with much art: the outſide conſiſts of coarſe ſtraw, and ſtiff ſtalks of herbs; the inſide of ſofter materials: the ſhape reſembles that of a retort: it is ſuſpended from

* Certhia Cærulea, *Linn. & Gmel.*
Certhia Cayanenſis Cærulea, *Briſſ.*
The Certhia of Guiana, *Bancroft.*
The Blue Creeper, *Edw. & Lath.*

Specific character: "It is ſky blue; a bar on its eyes; its throat, the quills of its wings and tail, black."

the

the end of a pliant branch, and the aperture faces the ground. The bird enters the neck, and creeps into the belly of the retort, which is its proper neſt. By this contrivance, the hatch is guarded againſt the viſits of ſpiders, lizards, and other intruders. Wherever weak animals ſubſiſt, unprotected by man, we may infer that they are induſtrious.

The author of the Eſſay on the Natural Hiſtory of Guiana * mentions a bird very ſimilar to the preceding, only its tail is of an uncommon length. Muſt we reckon this a male in its full perfection, or another variety of the ſame ſpecies?

* Bancroft. T.

II. THE

II.

THE BLACK-HEADED GREEN AND BLUE GUIT-GUIT, *Buff.*

Certhia-Spiza, *Linn. & Gmel.*
Certhia Americana Viridis Atricapilla, *Briss.*
Avicula Americana Altera, *Seba.*
The Black-headed Creeper, *Lath.*

THE plumage of this American bird consists of three or four colours, which are disposed in distinct masses, without any intermixture or shading: a velvet black on the throat and head only; deep blue under the body; bright green on all the upper side, including the tail and the wings; but the tail is of a deeper shade: the inferior coverts of the wings are cinereous brown, edged with green, and the bill is whitish.

Total length, five inches and a quarter; the bill, nine lines; the *tarsus*, the same length; the middle toe seven lines, rather longer than the hind toe; the tail eighteen lines, consisting of twelve equal quills, and exceeding the wings eight or ten lines; the alar extent unknown.

It is nearly as large as the chaffinch. We are not certain in what part of America it occurs:

curs: but moſt probably it inhabits the ſame regions with the two preceding. [A]

[A] Specific character of the *Certhia-Spiza*: "It is green; its head and wing quills are blackiſh."

VARIETIES of the BLACK-HEADED GREEN AND BLUE GUIT-GUIT.

* 1. THE BLACK-HEADED GREEN GUIT-GUIT. The head is black, as in the preceding, but not the throat: it is of a beautiful green, as are all the upper and under ſides of the body, including the ſuperior coverts of the wings: their quills are blackiſh, and alſo thoſe of the tail, but all bordered with green, the only colour that appears when the parts are at reſt: the inferior coverts of the wings are brown cinereous, bordered alſo with green: the bill is yellowiſh at its baſe, blackiſh above,

* Certhia-Spiza, var. *Lath. Ind.*
Certhia Braſilienſis Viridis Atricapilla, *Briſſ.*
Sylvia Viridis Capite Nigro, *Klein.*
The Green Black-capped Flycatcher, *Edw.*
The Black-capped Creeper, *Lath. Syn.*

whitiſh

whitiſh below, and the legs are of the ſame deep lead colour. The relative dimenſions are the ſame as in the preceding bird, only the tail is rather longer, and exceeds the wings eleven lines; the alar extent is ſeven inches and an half.

*2. The WHITE-THROATED GREEN AND BLUE GUIT-GUIT. The blue is ſpread on the head, and the ſmall ſuperior coverts of the wings: the throat is white: all the reſt of the plumage the ſame as in the preceding variety, except that in general the green is uniformly lighter, and on the breaſt are ſcattered a few ſpots of a deeper green: the bill is blackiſh above, white below, according to Briſſon; and, on the contrary, whitiſh above, and deep cinereous below, according to Edwards: the legs are yellowiſh.

With regard to the dimenſions, they are preciſely the ſame as in the preceding bird; and this conformity has made Edwards ſuſpect, that the two belong to the ſame ſpecies.

* Certhia-Spiza, var. 2. *Lath. Ind.*
Certhia Braſilienſis Viridis, *Briſſ.*
Sylvia Viridis Capite Cyaneo, *Klein.*
The Blue-headed Green Flycatcher, *Edw.*
The Blue-headed Creeper, *Lath. Syn.*

3. THE

3. THE ALL-GREEN GUIT-GUIT. All the under ſide of the body is deep green, tinged with blueiſh, except the rump, which, as well as the throat, and the under ſide of the body, is of a lighter green, tinged with yellowiſh: the brown of the wings is here black: the bill and legs are blackiſh; but there is a little fleſh colour near the baſe of the lower mandible.

This bird is found in Cayenne, and in Spaniſh America: it is of the ſame ſize with the preceding, and nearly the ſame proportions, except that the bill is rather ſhorter, and more ſimilar to that of the ſugar-birds.

III.

THE SPOTTED GREEN GUIT-GUIT.

Certhia Cayana, *Linn. Gmel. & Briſſ.*
The Cayenne Creeper, *Lath.*

THIS bird is ſmaller than the green guit-guits which we have juſt deſcribed, and it is differently proportioned. The upper ſide of the head and body of a fine green, though ſome-

* Certhia Spiza, var. 3. *Lath. Ind.*
The All-Green Creeper, *Edw. & Lath.*

what

4

what brown (variegated with blue in ſome ſubjects): on the throat is a mark of light rufous, encloſed on both ſides by two blue bars, which are very narrow, and accompany the lower mandible: the cheeks are variegated with green and whitiſh; the breaſt and the under ſide of the body marked with ſmall ſtreaks of three different colours, ſome blue *, others green, and others white; the inferior coverts of the tail, yellowiſh; the intermediate quills, green; the lateral ones blackiſh, edged and terminated with green; the quills of the wings the ſame; the bill black; between the bill and the eye is a light rufous ſpot, and the legs are gray.

In the female, the colours are leſs decided, and the green of the upper ſide of the body is lighter: it has no ruſty caſt, neither on the throat, nor between the bill and the eye, and not a ſingle ſhade of blue in the whole of its plumage. I obſerved one in which the two bars that accompany the lower mandible, were green.

Total length, four inches and two lines;

* In the individual deſcribed by M. Koelreuter, there was no blue; but the throat was yellow, as well as the ſpace between the bill and the eye: I ſhould ſuppoſe it to be a young male, and not an adult female.

the

the bill, nine lines; the *tarſus*, ſix lines; the mid toe the ſame length, and exceeding that of the hind toe; the alar extent, ſix inches and three quarters; the tail fifteen lines, conſiſting of twelve equal quills, and projects five lines beyond the wings. [A]

[A] Specific character of the *Certhia Cayana*: "It is ſhining green; below ſtriped with white; its tail quills green; the lateral ones blackiſh within."

IV.

THE VARIEGATED GUIT-GUIT, *Buff.*

Certhia Variegata, *Gmel.*
Certhia Americana Varia, *Briſſ.*
Sylvia Verſicolor, *Klein.*
The Variegated Creeper, *Lath.*

NATURE ſeems to have taken pleaſure in decorating the plumage of this bird with variety and choice of colours: bright red on the top of the head; fine blue on the back of the head; blue and white on the cheeks; two ſhades of yellow on the throat, the breaſt, and all the under ſide of the body; yellow, blue, white, blackiſh, on the upper ſide of the body, including the wings, the tail and their ſuperior

 coverts,

coverts. It is ſaid to be an American bird; but the part of that continent which it commonly inhabits is not aſſigned. It is nearly as large as the chaffinch.

Total length, five inches; the bill, nine lines; the *tarſus*, ſix lines; the mid toe ſeven lines, rather longer than the hind toe; the nails pretty long; the tail ſeventeen lines, and exceeds the wings five or ſix lines. [A]

[A] Specific character of the *Certhia Variegata*: "It is variegated with blue, black, yellow and white; below ſaffron; its top red; the back of its head blue."

V.

THE BLACK AND VIOLET GUIT-GUIT, *Buff.*

Certhia Braſiliana, *Gmel.*
Certhia Braſilienſis Violacea, *Briſſ.*
The Black and Violet Creeper, *Lath.*

THE throat and fore ſide of the neck are of a ſhining violet; the lower part of the back, the ſuperior coverts of the tail, and the ſmall ones of the wings, are violet, bordering on ſteel

ſteel colour; the upper part of the neck and back, of a fine velvet black; the belly, the lower coverts of the tail and of the wings, and the great ſuperior coverts of the wings, of a dull black; the top of the head, of a fine gold green; the breaſt, purple cheſnut; the bill blackiſh, and the legs brown. This bird is found in Brazil: it is of the ſize of the crowned wren.

Total length, three inches five lines; the bill, ſeven lines; the *tarſus*, five lines and a half; the mid toe five lines, rather longer than the hind one; the alar extent, four inches and a quarter; the tail thirteen lines and a half, conſiſting of twelve equal quills, and exceeds the wings five or ſix lines. [A]

[A] Specific character of the *Certhia Braſiliana*: "It is black; its top, gold green; its rump and its throat, violet; its breaſt, bay purple."

VI.

THE SUGAR-BIRD.

Le Sucrier, *Buff.*
Certhia Flaveola, *Linn. & Gmel.*
Certhia, ſeu Saccharivora Jamaicenſis, *Briſſ.*
The Black and Yellow Creeper, *Lath.*

ITS ordinary food is the ſweet viſcous juice of the ſugar cane, which it ſucks through the cracks of the ſtalk: ſo I have been informed by a traveller who reſided many years at Cayenne. In this reſpect it reſembles the humming birds; it is alſo exceedingly ſmall; and from the relative length of its wings, it approaches that of Cayenne, though it differs by the length of its legs and the ſhortneſs of its tail. I ſuſpect that the ſugar-birds likewiſe eat inſects, though this has not been mentioned.

In a male from Jamaica, the throat, the neck, and the upper ſide of the head and body, were of a fine black, but with ſome exceptions; for there were white eyebrows, traces of white on the great quills of the wings, from their origin to more than half their length, and alſo on the

the tips of all the lateral quills of the tail; the edge of the wings, the rump, the flanks and the belly, of a fine yellow, which ſpreads and grows dilute on the lower belly, and becomes whitiſh on the inferior coverts of the tail.

The ſpecies is diffuſed through Martinico, Cayenne, and St. Domingo, &c. but the plumage varies a little in theſe different iſlands, though nearly in the ſame parallel. In the ſugar-bird of Cayenne*, the head is blackiſh; there are two white eyebrows, which extending meet behind the neck: the throat is light aſh gray; the back and the ſuperior coverts of the wings of a deeper aſh gray; the quills of the wings and of the tail aſh gray, bordered with cinereous; the anterior part of the wings bordered with lemon yellow; the rump yellow; the breaſt and the under ſide of the body alſo yellow; but this colour is mixed with gray on the lower belly: the bill is black, and the legs blueiſh: the tail projects a very little beyond the wings.

This bird has a very delicate note, *zi*, *zi*, and, like the humming bird, alſo ſucks the juice of plants. Though I have been ſtrongly aſſured that the one which I have deſcribed is a male,

* The negroes and creoles of St. Domingo call it *Sicouri*.

I muſt

I muſt own that it bears great reſemblance to the female from Jamaica*, only this has a whitiſh throat, and a cinereous caſt where the other was blackiſh; the eyelids yellowiſh white; the anterior part of the wings edged with white, and the rump of the ſame colour with the back; the five pairs of lateral feathers of the tail terminated with white, according to Edwards; the ſingle exterior pair, according to Briſſon: laſtly, the greateſt quills of the wings white, from their origin more than half their length, as in the male.

Sloane ſays, that this bird has a very ſhort but pleaſant warble; but that was probably the female, and the male would ſtill have a ſweeter air. The ſame obſerver, who diſſected one of theſe birds, tells us that the throat and gizzard were ſmall, the latter not muſcular, and lined by an inadheſive membrane; the liver bright red, and the inteſtines rolled into many circumvolutions.

I have ſeen a ſugar-bird from St. Domingo, in which the bill and the tail were rather ſhorter, the eyebrows white, and on the throat a ſort of whitiſh mark, which was larger than in the

* Certhia Flaveola, var. 1. *Linn. & Gmel.*
Certhia Martinicana, ſeu Saccharivora, *Briſſ.*
The Yellow-bellied Creeper, *Edw.*

above

above female; in all other reſpects it was exactly ſimilar.

Laſtly, Linnæus regards the Bahama creeper of Briſſon as the ſame with the ſugar-birds of Martinico and Jamaica*. Its plumage is indeed very ſimilar; all the upper ſide is brown, including even the quills of the wings and of the tail; the latter are whitiſh beneath; the throat is light yellow; the anterior edge of the wings, their inferior coverts, and the reſt of the under ſide of the body, of a deeper yellow as far as the lower belly, which is of the ſame brown as the back. Further, this bird is larger than the other ſugar-birds; ſo that it may be regarded as a variety of ſize and even of climate. The following are the dimenſions compared:

	Bahama Sugar Bird,			Jamaica Sugar Bird	
	Inches.	Lines.		Inches.	Lines.
Total length - - - -	4	8	—	3	7
Do. not including the tail -	0	32	—	0	27
The bill - - - - - - - -	0	6	—	0	6
The *tarſus* - - - - - -	0	6½	—	0	7
The middle toe - - - -	0	5½	—	0	6
The hind toe - - - - -	0	5 and more		0	4 or 5
The alar extent - - - - -	7	0	—	unknown	
The tail conſiſting of 12 quills	2	0	—	1	4
Its exceſs above the wings -	0	15 or 16		0	5 or 6

* Certhia Flaveola, var. 2. *Linn. & Gmel.*
The Bahama Titmouſe, *Cateſby.*

 The

The name *luſcinia,* which Klein beſtows on it, ſhews that he regarded it as a ſinging bird; another point of analogy to the Jamaica ſugar-bird. [A]

[A] Specific character of the *Certhia Flaveola*: "It is black, below yellow; its eyebrows partly white; its outermoſt tail quills tipt with white"

END OF THE FIFTH VOLUME.

Printed in the United States
By Bookmasters